Evolution by the Numbers

Rhetoric of Science and Technology
Series Editor: Alan G. Gross

The rhetoric of science and technology is a branch of rhetorical criticism that has grown rapidly since its inception four decades ago. Its initial focus was the texts of such well-known scientists as Darwin, Newton, and Watson and Crick. The field has since expanded to encompass important work on interdisciplinarity, the role of rhetorical schemes, the popular meanings of the gene, the rhetorical history of the scientific article, the question of incommensurability, and the critical engagement with emergent technologies. But this work and these topics by no means exhaust the field. Although the point has already been made that science and technology are in some sense rhetorical, the field remains open to new topics and innovative approaches. For submission information, please visit the series page at http://www.parlorpress.com/science.

Other Books in the Series

Communicating Science: The Scientific Article from the 17th Century to the Present by Alan G. Gross, Joseph E. Harmon, and Michael S. Reidy (2009)

EVOLUTION BY THE NUMBERS

The Origins of Mathematical Argument in Biology

James Wynn

Parlor Press
Anderson, South Carolina
www.parlorpress.com

Parlor Press LLC, Anderson, South Carolina, USA

Printed in the United States of America

S A N: 2 5 4 - 8 8 7 9

Library of Congress Cataloging-in-Publication Data

Wynn, James, 1972-
Evolution by the numbers : the origins of mathematical argument in biology / James Wynn.
p. cm. -- (Rhetoric of science and technology)
Includes bibliographical references and index.
ISBN 978-1-60235-216-2 (pbk. : alk. paper) -- ISBN 978-1-60235-217-9 (hardcover : alk. paper) -- ISBN 978-1-60235-218-6 (Adobe ebook) -- ISBN 978-1-60235-219-3 (epub)
1. Evolution (Biology)--Statistical methods. 2. Evolution (Biology)--Mathematics. 3. Biometry. I. Title.
QH362.W96 2011
576.801'51--dc22

2011010840

Cover design by Michael West, Cygnus Design.
Author photo by Shane Wynn of Shane Wynn Photography.
Printed on acid-free paper.

Parlor Press, LLC is an independent publisher of scholarly and trade titles in print and multimedia formats. This book is available in paper, cloth and eBook formats from Parlor Press on the World Wide Web at http://www.parlorpress.com or through online and brick-and-mortar bookstores. For submission information or to find out about Parlor Press publications, write to Parlor Press, 3015 Brackenberry Drive, Anderson, South Carolina, 29621, or email editor@parlorpress.com.

EVOLUTION BY THE NUMBERS

The Origins of Mathematical Argument in Biology

James Wynn

Parlor Press
Anderson, South Carolina
www.parlorpress.com

Parlor Press LLC, Anderson, South Carolina, USA

Printed in the United States of America

S A N: 2 5 4 - 8 8 7 9

Library of Congress Cataloging-in-Publication Data

Wynn, James, 1972-
Evolution by the numbers : the origins of mathematical argument in biology / James Wynn.
p. cm. -- (Rhetoric of science and technology)
Includes bibliographical references and index.
ISBN 978-1-60235-216-2 (pbk. : alk. paper) -- ISBN 978-1-60235-217-9 (hardcover : alk. paper) -- ISBN 978-1-60235-218-6 (Adobe ebook) -- ISBN 978-1-60235-219-3 (epub)
1. Evolution (Biology)--Statistical methods. 2. Evolution (Biology)--Mathematics. 3. Biometry. I. Title.
QH362.W96 2011
576.801'51--dc22

2011010840

Cover design by Michael West, Cygnus Design.
Author photo by Shane Wynn of Shane Wynn Photography.
Printed on acid-free paper.

Parlor Press, LLC is an independent publisher of scholarly and trade titles in print and multimedia formats. This book is available in paper, cloth and eBook formats from Parlor Press on the World Wide Web at http://www.parlorpress.com or through online and brick-and-mortar bookstores. For submission information or to find out about Parlor Press publications, write to Parlor Press, 3015 Brackenberry Drive, Anderson, South Carolina, 29621, or email editor@parlorpress.com.

Acknowledgments

This book would not be possible without the help and generosity of many people and organizations, a few of whom I would like to briefly thank. Travel to the archives at the University College London and the John Innes Horticultural Institute was made possible through the generosity of Carnegie Mellon's Falk Grant and Berkman Faculty Development Fund. I would also like to thank the University College London and John Innes Horticultural Institute, along with the Hunt Botanical Library for opening up their collections for my research. Additionally, the University of Adelaide, The Royal Society of Edinburgh, Cambridge University Press, and The Royal Society of London deserve recognition and thanks for the use of the figures and tables in this text.

In addition to material resources, this book has benefitted from the intellectual generosity of a number of scholars who have shared their time and expertise with me. I am particularly indebted to Jeanne Fahnestock, who has read and critiqued this book in many of its different stages. I am also grateful to Alan Gross for his incisive editing of the text. In addition, I would like to thank Lorain Daston for taking time to read and provide feedback on the historical framing discussed in the second chapter of the book. Colleagues who have read and offered advice on various parts of this text also deserve recognition, including: Leah Ceccarelli, Andreea Ritivoi, Kristina Straub, Lynda Walsh, and Michael Witmore.

Finally, the emotional support of my wife Gina has been crucial to the completion of this project. I would like to express my gratitude to her for all her patience and personal sacrifice.

Some of the material in this book draws on material in journal articles previously published by the author. The material from these articles has been modified and revised. Chapter 3 was developed from two papers. The first, "Arithmetic of the Species: Darwin and the Role of Mathematics in His Argumentation," appeared in *Rhetorica* 27.1

(2009): 76–100. The material in this paper has been considerably revised, and much new material has been added. The second, "A New Species of Argument: Darwin and the Role of Mathematics in His Argumentation," appeared in *19th Century Prose* 38.1 (2011). The material in this paper has been revised, and new material has been added. Chapter 4 is based in part on "Alone in the Garden: How Gregor Mendel's Inattention to Audience May Have Affected the Reception of his Theory of Inheritance in 'Experiments in Plant Hybridization,'" which appeared in *Written Communication* 24.1 (2007): 3–27. The material in this paper has been considerably revised, and much new material has been added.

Foreword

Variation, Evolution, Heredity, and Mathematics in the 21st Century

All three non-adaptive forces of evolution—mutation, recombination, and random drift—are stochastic... and can generally only be understood in probabilistic terms. It is well-known that most biologists abhor things mathematical, but the quantitative details really do matter.

—Michael Lynch

In the twenty-first century, the concepts of variation, evolution, and heredity have influenced science, technology, and the public imagination in ways that could never have been imagined by their developers. Evolution has been embraced by the public to explain complex transformations in everything from organisms to economies, and in the process, has divided the public spheres on issues as diverse as religion, science, and public policy. Variation and heredity have similarly become part of our modern social and cultural awareness. As our capabilities to modify genes in plants and animals grow, so do the difficulties of our deliberations over whether and to what extent we should bioengineer our way to a better world.

Though we easily recognize how these ideas influence our social and cultural landscape, most of us rarely consider how they transform science. This task falls to historians, philosophers, and sociologist of science. However, even scholars in these fields have not considered all of the consequences of these notions. One of the effects that has not been explored is the impact of these ideas on the development of *argument* in the biological sciences. This book examines how the concepts of variation, evolution, and heredity, introduced by Charles Darwin and Gregor Mendel, transformed argument in the biological sciences by encouraging the growth of mathematical argumentation.

Mathematics and Modern Investigations of Variation, Evolution, and Heredity

Unlike scientific ideas, which regularly filter into the public's awareness, mathematical aspects of scientific argument tend to develop quietly and anonymously. Despite their low profile, they are ubiquitous and in modern investigations of variation, evolution, and heredity. By going behind the scenes of current research in these fields, it is possible to illustrate just how important they are.

The extent to which these research fields have come to rely on mathematics is nothing short of extraordinary. At the dawn of the twentieth century, very few researchers investigating these phenomena would have been employing mathematical methods or arguments. However, in the twenty-first century, these methods pervade their work. This pervasiveness is evidenced by the spectacular growth in the last fifty years of mathematical fields of study related to these phenomena, including: population genetics, molecular genetics, biostatistics, bioinformatics, computational biology, and quantitative genetics. The ubiquity of mathematics is also evident in the range of subjects that are being examined using quantitative methods. According to Alan Templeton, a professor of population genetics at Washington University in St. Louis, mathematical models are currently being used in a variety of research areas, including "wildlife conservation projects, research assessing what it means to be human, and investigations tracking the historical development of disease."

One publicly salient application of mathematics to the study of variation, evolution, and heredity has been the use of DNA to track the historical migrations of human populations as they spread out of Africa. This subject has been the focus of attention in a number of works in the popular media, including books such as Steven Olson's *Mapping Human History* and Brian Sykes's *The Seven Daughters of Eve,* websites like Wikipedia's "Human Evolutionary Genetics," and televised specials like PBS's *Journey of Man.* In all of these media, however, the role of mathematics in the science is invisible. Closer scrutiny of these popularizations, though, offers a sense of the true extent to which mathematics contributes to the science that captures the public's imagination.

In the television documentary *Journey of Man,* for example, English geneticist Dr. Spencer Wells travels the world tracing the hereditary path of our human ancestry by following the physical route by which it migrated out of Africa. As is the case for most popularizations of science, the main focus of the documentary is on the human story. Though it gets second billing, science does appear throughout the documentary. Before Wells leaves on his trip, for example, he visits his geneticist mentor, Luca Cavalli-Sforza, to talk about the foundations of research into human genetic variation. He also takes breaks in his travels to explain key scientific points, such as what scientists know about gene change over time and how it helps them establish relationships between modern humans and their progenitors.

However, the mathematical work that makes identifying these relationships possible receives only the briefest of acknowledgements. Interspersed throughout the documentary are visual images of peaked line graphs on computer screens. In addition, Wells makes brief reference to "the clear data" that has sent him to Africa in search of the Kalahari bushman whose genetic heritage represents the starting point of the human journey of geographic expansion and genetic diversification. The obliqueness and briefness with which the documentary treats the contributions of mathematical argument creates the impression that it played almost no substantive role in understanding the spread of our ancestors. In reality, however, Wells's trip would not have had a scientifically supportable itinerary without quantitative data and mathematical methods for managing, comparing, and analyzing that data.

For example, to establish the chain of genetic ancestry from fixed mutations in the Y chromosome—which Wells uses as the scientific basis for his travels—thousands of blood samples taken in the field would first have to be processed. In the initial phases, chemical and physical procedures would be used to extract and precipitate DNA. Once the DNA had been extracted, it would be "unzipped," bonded to other known bits of DNA, and run through a process of electrophoresis where it would be separated out and read by a laser.

Once the DNA was scanned and identified, mathematics would take on its essential role in the science. The information, read by laser from the DNA, would be stored in a database whose architecture would not be possible without the use of complex mathematical algorithms. Then this information would be compared to other samples in large databases, again using sophisticated algorithms. To determine the general relatedness of the sample of DNA to a population group, researchers would apply formulae to calculate the probability of the DNA's belonging to a particular group based on the absence or presence of certain genetic markers. Finally, scientists would establish the place, say for a Kalahari bushman's Y chromosome, in the larger sequence of genetic diversification amongst human population with another set of formulae. These formulae would be used to detect the absence or presence of key mutations in the bushman genome and to compare them to the mutations present or absent in other human populations.

Because of long-term efforts to gather genetic data and improve the speed of its analysis, scientists now have more information relevant to investigating variation, evolution, and heredity than ever before. The extent of this data and the type of inquiries it supports means that research such as the kind popularized by Wells cannot be conducted without mathematics. Its necessity is evidenced by the emergence and coalescence of a number of mathematical subfields in modern biology—such as bioinformatics, molecular genetics, population genetics, biostatistics, and statistical genetics—dedicated to meeting the needs of a quantitative science (Templeton). Researchers in bioinformatics, for example, devote their efforts to developing databases, algorithms, and statistical and computational techniques for analyzing and managing massive data sets. With the complete sequencing of the human genome and other important organisms, the amount of genetic data that needs to be organized and synthesized has grown. The human genome, for example, has between twenty and twenty-five thousand genes and other functional elements, with an estimated three billion base pairs. To collect this data set, the institutions working on the Human Genome Project collaboratively sequenced genes for fifteen years. Computer scientists in bioinformatics employ their mathematical skills to develop more powerful algorithms for ensuring that data on this scale can be properly stored and retrieved for scientific research.

Whereas some biomathematical researchers devote their talents to managing data, others use their mathematical skills to develop formulae to pose and solve important questions about variation, evolution, and heredity, such as how closely species are related and how diseases have emerged and developed over time. Molecular geneticists, for example, might test hypotheses about the degree of relatedness between organisms by developing genetic taxonomies or gene trees. These trees require special algorithms designed to calculate the proximity of organisms to one another based on their genetic divergence in some physical trait. For example, a molecular geneticist might compare the order of amino acids in the red blood cells of humans, pigs, mice, and chickens. Using one or more of a handful of standard mathematical methods for calculating relatedness between organisms, he/she would conclude that, evolutionarily speaking, humans are closer to pigs than chickens (Hartl and Jones 612–13).With these methods, molecular geneticists are beginning to provide better insight into relations of descent between organisms, including ones that would have likely eluded qualitative taxonomists, such as the water lily's (*Nuphar polysepalum*) position as the genetic progenitor of the oak tree and all other seed-bearing plants (National Science Foundation).

Along with molecular biologists, population geneticists also use established mathematical algorithms to describe changes in organic populations. They rely, for example, on the algebraic Hardy-Weinberg principle as a model for the distribution of genes in a population under random mating conditions. In this endeavor, mathematics plays a central role because it is used to define a hypothetical baseline for change in the rate of alleles (the different possible gene types at a specific location on a chromosome) in a population against which the effects of natural selection, population size, mutation, migration, and random drift can be assessed. Calculations like these are essential to a number of modern applications, such as scientific breeding programs, assessments of the efficacy of screening for genetic disease factors, and the estimation of biodiversity.

Finally, a discussion of the importance of mathematics to modern investigations of variation, evolution, and heredity would be incomplete without mentioning the general value of statistics and probability in the day-to-day pursuit of scientific research. In modern biological research, investigations regularly begin and end with statisticians or biostatisticians carefully assessing the methods and results of experiments. Trained in statistics and probability, these members of a research team provide guidance to laboratory scientists on how to structure their experiments so that they limit the influence of factors which might bias their outcomes. For example, a lab's biostatistician might advise geneticists working in disease research on techniques for random sampling to ensure that they have a data set from the general population for a genetic trait that might be used comparatively to identify genetic disease markers in a population of interest. After the tests are run and the data are collected, statisticians and biometricians are also tasked with calculating the reliability of the results and assessing the data to determine whether, if any, significant patterns emerge. These duties are so important to modern genetic research that Eleanor Feingold, a quantitative geneticist at the University of Pittsburgh, explained, "a lab of any reasonable size would have a biostatistician, a quantitative geneticist, or a statistician attached to it."

A behind-the-scenes investigation of operations of modern research into variation, evolution, and heredity reveals: (1) that these phenomena cannot reasonably be investigated without mathematics and (2) that because of the increasing size and availability of data on these phenomena, the importance of mathematics will continue to grow. For these reasons, understanding both the role of mathematical argument in science and how that role came to be established, which are the subjects of this book, should be considered important topics of exploration.

Rhetoric, Mathematics, and Science

Although modern research in variation, evolution, and heredity would be impossible without mathematics, there was a time when these phenomena were explored largely without it. The focus of this book is the one hundred-year period between the publication of *The Origin of Species* and the emergence of modern programs of population and quantitative genetics in the nineteen fifties and sixties. During this critical period of development, mathematics and its capacity to generate reliable knowledge about organic populations was disputed. The goal of this text is to explore some of the reasons why mathematical argument was resisted in these early periods, and how it was advocated for either successfully or unsuccessfully by natural researchers who wanted to advance its credibility and explore the possibilities for its use.

To examine the use of and debates about mathematics in this formative period, this investigation turns to the methods and tools of *rhetoric,* a field of research and analysis devoted to the study of human communication, argument, and persuasion. With the aid of concepts and methods from this field, the book examines choices in language, organization, and argument in discourse located within specific social, epistemological, and cultural/historical contexts. Examining these dimensions of discourse in context permits characterizations of the goals and beliefs of arguers, the perceptions they have of their audiences, and the suitability of their choices in argument and communication. By investigating these facets of argument and persuasion, this book aims to better understand mathematical argument in a scientific context as well as explore what this relationship reveals about the practical value of rhetorical tools and concepts in understanding it.

Although the text is written primarily with philosophers, historians, sociologists, and rhetoricians of science in mind, every effort has been made to accommodate a broader educated audience of readers. Non-specialist readers who follow the subjects of mathematics, genetics, and evolution will likely find their interests reflected in the choice of topics and figures being investigated in this book. Well-known researchers such as Darwin and Mendel will be discussed, and fresh perspectives on their work as mathematical argument will be examined. Chapter 3, for example, explores in detail not only Mendel's mathematical arguments in his famous paper, "Experiments in Plant Hybridization," but also the historical context in which he makes these arguments. Assessing these dimensions of Mendel's work reveals his reliance on the mathematics of probability as a source of invention for his pea experiments as well as his overconfidence that by using mathematical arguments he could persuade his audience to accept the general validity of his hereditary law.

Chapter 2 looks at Darwin's work from a seldom–examined, mathematical perspective and reveals the extent to which the self-proclaimed mathematical bumbler relied on quantitative evidence and arithmetically informed arguments to invent and support some of his central conclusions in *The Origin of Species.* An examination of Darwin's letters, diaries, his "big species" book (the original manuscript from which *The Origin of Species* was abstracted), and his arguments in *The Origin of Species* reveal that Darwin hoped to place biology on par with the physical sciences by giving it a solid, mathematical foundation. Other chapters in the book are devoted to less-well-known—but no less important or interesting—figures such as Francis Galton, Karl Pearson, and R.A. Fisher, all of whom play important roles in the development of a mathematical science of variation, evolution, and heredity.

In addition to persons of interest, this book also explores topics in science and mathematics with broad appeal. For example, it engages with the perennial issue of how scientific knowledge is validated. By examining the successes and failures of the scientists featured in the text, it suggests that when a scientific paradigm cannot be relied on to establish the appropriateness of mathematical arguments, arguers can turn to beliefs and values outside of science. Further, it explores the question. "How reliable and successful are mathematics in describing real phenomena?" This query is central to our current public and scientific concerns as we turn to polls, statistics, and probabilistic assessments with increasing frequency to make decisions about products, cures, risks, and candidates. What this investigation reveals is that nineteenth and early twentieth century scientists approached new mathematical methods and their conclusions with a healthy dose of skepticism. They offered legitimate resistance to mathematically informed theories of variation, evolution, and heredity that had insufficient evidence or blatantly disregarded important aspects of biological phenomena. In a few instances, however, their rejections were premature and prejudiced by either their ignorance of mathematics or their personal commitments to other methods of analysis.

Whether you read this book for the characters or the concepts, the goal is the same: to scrutinize, using the tools of rhetoric, the texts, arguments, and contexts involved in the development of the relationship between biological investigations of variation, evolution, and heredity and mathematics from the middle of the nineteenth to the beginning of the twentieth century. In investigating these phenomena, it endeavors to show how Darwin's and Mendel's ideas about them influenced the transformation of biology from a predominantly qualitative science to one with a vital, mathematical component. It also reveals how difficult this transformation—which we now take for granted as the very essence of our modern sciences of genetics and evolution—really was. By approaching science from a rhetorical perspective as a process of argument and deliberation rather than a product (as it is so often presented in the popular media), we can develop a greater appreciation both for the value of mathematics as a source for knowledge about nature and for the difficult and sometimes circuitous path by which that confidence is earned.

Evolution by the Numbers

1 Introduction

I assert . . . that in any special doctrine of nature there can be only as much proper science as there is mathematics therein. For . . . proper science, and above all proper natural science, requires a pure part lying at the basis of the empirical part.

—Immanuel Kant

In the twentieth and twenty-first centuries, there has been a substantial expansion in the number of fields applying mathematics to their investigations of natural and social phenomena. Areas of social research such as psychology and sociology, which had traditionally been qualitative, have developed robust quantitative components, such as standardized intelligence tests and statistical surveys to assess the habits, beliefs, and practices of populations. In addition, fields of natural investigation. particularly genetics and evolutionary biology, have expanded their methods as described in the Forward from observation and experiment to include mathematical descriptions of the genome and the change and distribution of variation within organic populations. As a result of this expansion, mathematics has become a ubiquitous aspect of what makes a discipline "scientific," making Kant's invocation that "there can only be as much proper science as there is mathematics therein" seem even more relevant today as it was when he wrote it in 1786.[1]

Although this mathematical expansion has helped us better understand social, psychological, and biological phenomena, the path to developing and adopting mathematics in science has not always straight or easy. One widely-cited example of a mathematical science with a turbulent beginning is population genetics. In the twenty- first century, this vital field of genetics research has its own textbooks, specialists, and places in the academy. Its recent success, however, obscures a less flourishing past. Although the basic scientific and mathematical foundations for population genetics were largely in place by the first decade of the twentieth century, almost thirty years elapsed before it garnered sufficient attention and support to establish itself as a field of research worthy of an individuated identity.

Historians interested in the intellectual foundations of population genetics and its transition into an important field of inquiry have reached back to Darwin and traced its development forward, hoping to understand the reasons why such a productive, modern field of study had such a difficult maturation. Perhaps the most well-known investigation into this mystery was undertaken by historian William Provine, whose groundbreaking book, *The Origins of Theoretical Population Genetics*, suggests that the turmoil associated with the rise of the field was largely the consequence of an ideological conflict between Darwinians and Mendelians (*ix-x*).

According to Provine, Darwin and his later followers, the biometricians, believed in *continuous* variation in which differences between members of a species arose by the slow accretion of small variations over long periods of time. Mendelians and the supporters of mutation theory, on the other hand, believed that variation was *discontinuous*: varieties appeared suddenly and could introduce dramatic changes into individuals and populations of organisms. By tracing these notions about variation from Darwin through the debates between the biometricians and the Mendelians, Provine concludes that it was only when the differences between these ideological positions were resolved in the work of R.A. Fisher, J.B.S. Haldane, and Sewall Wright, that a research field of population genetics emerged (Provine 131).

Although Provine and other historians rightfully devote attention to how ideological conflicts over variation complicated the development of population genetics, their focus excludes other elements that could have contributed significantly to population genetics' "torturous" development (Provine *ix*). One important element that has not been considered is whether and to what degree the beliefs about the acceptability of using mathematics to make arguments about biological phenomena might have contributed to the difficulties in establishing the field.

Although it is difficult from a twenty-first century perspective to image mathematics not being a legitimate means of researching variation, evolution, and heredity, scrutiny of the work of early researchers such Charles Darwin, Gregor Mendel, Francis Galton, Karl Pearson, and R.A Fisher suggests that this has not always been the case. Attention to their work and its reception reveals that mathematical approaches to these phenomena were caught up in a cycle of development, conflict, and persuasion that lasted almost one hundred years, a cycle that has all but been forgotten as science looks to the future and eviscerates from memory the useless, blind allies and conflicts that led to its current position. In hoping to understand the development of scientific knowledge, however, we need to look at the process of making knowledge, not just the results. Investigating these long-forgotten conflicts can help us understand not only what people believed, but also how they were moved to change their beliefs, what they perceived were good reasons for accepting or rejecting a particular position, and what lines of argument dominated the scientific landscape.

Despite the importance of mathematics to scientific argument and epistemology, there have been few historical-philosophical, sociological, or rhetorical investigations of how scientists argue with or about the use of mathematics.[2] In the history of science, the intersection between argument, science, and mathematics has been investigated by historian Peter Dear, whose book, *Discipline and Experience: The Mathematical Way in the Scientific Revolution*, examines sixteenth, seventeenth, and eighteenth century disputes among natural philosophers about whether mathematics could serve legitimately and authoritatively as a source for arguments about nature. In the book, Dear attempts to make sense—in the context of eighteenth century natural philosophy—of both the novelty of Newton's physico-mathematical argument strategy and its importance to the development of a new paradigm for scientific research (248).

Although very few historians have examined the relationships between mathematics, science, and argument, there is evidence of a trend towards increasing attention to the subject. In a recent discussion, for example, in *Isis*—a top journal in the study of history and philosophy of science—titled, "Ten Problems in the History and Philosophy of Science," historian and philosopher Peter Galison lists a lack of understanding of the "Technologies of Argumentation" as problem number three for philosophers and historians of science. He raises the following questions for them to pursue:

> When the focus is on scientific practices (rather than discipline-specific scientific results *per se*), what are the concepts,

> tools, and procedures needed at a given time to construct an acceptable scientific argument? . . . Cutting across subdisciplines and even disciplines, what is the toolkit of argumentation and demonstration—and what is its historical trajectory? (116)

For rhetoricians of science, whose interests lay predominantly in the study of scientific argument and communicative practices, answering these sorts of questions about the relationship between mathematics, science, and argument would seem to be an important and fruitful undertaking. Despite the natural fit between scholarly interest and subject matter, however, very few rhetoricians have made efforts to examine the intersection between these three subjects. One notable exception is the work of Alan Gross, Joseph Harmon, and Michael Reidy in, *Communicating Science: The Scientific Article from the 17th Century to the Present.* In this book, the authors examine the developing conventions of argument and style in the scientific article, including brief descriptions of the use of mathematics.

Though the works of Dear, Gross, Harmon and Reidy begin a conversation about the role of mathematics in scientific argument, there are many important avenues currently unexplored. Questions—such as, "Do new mathematical methods have a different status of reliability as a source for arguments in science than existing ones?"; "If mathematical methods are not assumed *a priori* to be reliable, how do scientists make a case for their use in science?"; and "Can the reliability of mathematical methods and their use be debated and secured using methods outside of a framework of analytical argument?"— still remain and represent substantial lacunae in or understanding of the subject. The primary goal of this book is to explore these questions.

Just as scientists rely on rare diseases amd aphasias to understand the functioning of genes and language, this investigation turns its attention to a special case of the mathematization of a scientific field to find answers to these questions. Specifically, it examines the development of the mathematical study of variation, evolution, and heredity from the middle of nineteenth to the beginning of the twentieth century which eventually culminates in the emergence of important mathematical subfields of biology, including population genetics, quantitative genetics, and biostatistics. This development provides a unique opportunity to observe the nuances and difficulties in the relationships between mathematics, science, and argument.

By examining the conventions for arguing mathematically about natural phenomena and the successes and failures of advocates for a mathematical approach, I intend to advance four conclusions about the relationship of mathematics to scientific/biological argument:

- that novel mathematical arguments used to make claims about natural phenomena do not necessarily compel acceptance,
- that scientists arguing for novel mathematical warrants rely on a range of resources for generating good reasons to support their use,
- that arguments about and with mathematics in science can have non-analytical, rhetorical dimensions, and
- that conflicts over the appropriateness of using mathematics have complicated the development and acceptance of biomathematical fields such as population genetics.

A Rhetorical Approach to Scientific Epistemology

Any effort to discuss scientific argument and knowledge-making requires some explanation of one's philosophy of scientific epistemology. The epistemological perspective guiding this rhetorical investigation can be understood by contrasting it with positions on the subject that have been previously taken up by historians, philosophers, and sociologists of science.

In the last century, notions of scientific epistemology have tended either towards logical positivist/empiricist models of science or towards social constructivist models. For the logical positivist/empiricist, scientific propositions and theories are thought to be systematically verified or falsified by appealing to physical evidence in conjunction with logical-linguistic constructs and deductions. For the social constructionist, rationality is located in the commitments of a scientific community to seeing nature in a particular way. Although logical positivist/empiricist approaches to scientific epistemology were extremely influential in the late decades of the nineteenth and the early decades of the twentieth century, they met a series of challenges from philosophers such as Karl Popper, W.V. Quine, and Ludwig Wittgenstein, culminating in Thomas Kuhn's *The Structure of Scientific Revolutions*, which substantially decreased their appeal. Kuhn's investigation offered a fairly comprehensive vision of scientific knowledge-making that rejected the possibility that fixed, rational principles could be appealed to in times of epistemic crisis. From Kuhn's paradigmatic perspective, major changes in the conceptual framework of a scientific community could only occur when an existing paradigm had become so troublesome that its adherents began the process of developing alternative paradigms to replace it. This feature of paradigmatic change precludes falsification or verification by rejecting the possibility of a rational, external position from which a paradigm could be judged (145).

One of the consequences of Kuhn's model of scientific epistemology is that it not only eliminates the possibility for "objective" logical-linguistic constructs and deductions to guide argument and decision-making in science, but also the prospects for any reasonable common ground to exist between members of old and new paradigms. The absence of a third position, or alternative reasonable perspective, from which arguments supporting or challenging a paradigm over its alternative can be made or judged, raises important questions such as: "How is it that researchers working in a particular field during a time of revolution choose one paradigm over another?" and "How do communities of scholars with different points of view decide that one school of thought's natural metaphysic is sufficiently better than its competitors' and should be embraced as a paradigm?"

In response to the first question, Kuhn argues that the choice of a paradigm is made on the basis of a *personal* rather than a communal calculus. Novices entering a field with conflicting paradigms, for example, choose a paradigm to apprentice under according to their own individual sensibilities about which one they more closely identify with. Similarly, established participants in an existing paradigm either remain steadfast in their support for it, or experience a sudden, personal conversion to the alternative. In both cases, the transformation cannot be compelled by any commonly held good reasons for preferring one position over the other (Kuhn155). In response to the second question, Kuhn offers no criteria at all, stating only that "to be accepted as a paradigm, a theory must seem better than its competitors" (17).

As a challenge to logical positivist/empiricist approaches to scientific epistemology, Kuhn's concept of paradigm adoption swings away from what the rhetorical theorist Kenneth Burke, in the *Rhetoric of Motives*, calls a *grammatical stance:* a search for a set of *universal* propositions and procedures which would account for its epistemological robustness (21–23). In correcting the perceived errors of the grammatical position, however, Kuhn moves towards a radically opposed *symbolic stance* in which rationality is bound to *idiosyncratic,* personal reasons for choice rather than some loci of rationality shared by the larger community. The perspective on knowledge and argument employed in this investigation takes an epistemological middle ground between the universal and the idiosyncratic. This middle path is uniquely fitted to a rhetorical perspective because it rejects, on the one hand, analytical self-evidence by embracing the centrality of audience in argumentation, and in so doing, the necessarily communal and probabilistic nature of argument (Perelman and Olbrechts-Tyteca 1–10). On the other hand, it takes up the position that discourse communities overlap and interconnect and, as a consequence, good reasons can exist outside of a single discourse community and affect persuasion within it (Aristotle, *Rhetoric* I, ii 15–25). This allows for alternative avenues of rationality and common ground to exist, even in cases where different paradigms or schools of thought compete, and dispenses with the necessity of reverting to personal calculi to make decisions in these types of crises.

Rhetorical Method

From a rhetorical perspective, foci for analysis can include, but are certainly not limited to: (1) the good reasons and forms of evidence and argument that discourse communities find acceptable, (2) the effects of these dimensions of argument on the choices that speakers and writers make in constructing arguments, and (3) the ways that audiences judge their choices. Analyses centered on these foci address questions like, "Who is the audience for a scientific argument?"; "What conventions govern the way researchers participating in a particular scientific discourse community are expected to argue?"; "What facts, beliefs, and values do participants within a particular research community use to judge the validity and reliability of methods and conclusions?"; and "What broader sets of facts, beliefs, and values might influence the making or judging of arguments?"

Historical Analysis

This investigation relies on several different methods of analysis, including historical analysis, close textual analysis, and audience response analysis to draw conclusions about the relationship between scientific and mathematical arguments in the development of mathematical approaches to the study of variation, evolution, and heredity. Historical analysis is used to establish the dimensions of the scientific debates surrounding these phenomena and the perceived role of mathematics in the debate from the middle of the nineteenth to the beginning of the twentieth century. This aspect of analysis draws on a wide range of sources, including archived letters, scientific articles, philosophies of science, reader reviews of primary texts, and secondary historical accounts to establish the contours of the debate. These resources also supply evidence for characterizing the role of mathematical argument in science during the period under investigation.

Using philosophies of science to establish the conventions for scientific arguing in a particular historical period is not, to my knowledge, a method that has been previously exploited by rhetoricians of science. Current rhetorical work analyzing scientific argument relies either on a scientific figure's knowledge of rhetoric or on second-hand accounts of the conventions of scientific argument. In Jean Dietz Moss's work on the Copernican controversy, *Novelties in the Heavens,* for example, Moss offers readers historical evidence that scientific figures such as Galileo and Kepler had studied and/or taught rhetoric. This evidence proves a particular connection between rhetorical treatises and a scientist's use of characteristically rhetorical strategies of argument—such as the employment of ornamental language to gain the attention and admiration of the reader. While this approach offers a robust connection between specific rhetorical training and argument, it limits the rhetorical analyst to cases in which scientists can be proven to have had a rhetorical education. Though these limitations are not prohibitive in investigations of Renaissance science, they become severely restrictive for science in the nineteenth century—a time when rhetoric was largely absent from standard education and during which no new substantive treaties on rhetoric were published (Houlette ix). Further, by restricting rhetorical investigations to facets of argument learned through a rhetorical education, the range of scientific argument that can be explored is unnecessarily narrowed based on distinctions between dialectic and rhetoric, which are practically very difficult to maintain.

Other rhetorical analysts of science have adopted a broader sense of the argumentative territory open to rhetorical discussion and analysis. However, in their efforts to identify the conventions of scientific argument, they depend on secondary rather than primary sources. In Lawrence Prelli's substantive work on the rhetoric of science, *A Rhetoric of Science: Inventing Scientific Discourse,* for example, he relies on Thomas Kuhn's discussions in *The Structure* as the source for his *problem-solution topoi* and *evaluative topoi.*[3] Though Kuhn is certainly considered a reputable source for understanding scientific argument, neither he nor Prelli offer any primary source evidence to corroborate that scientists endorsed these lines of reasoning as conventional places for finding arguments in science.

The methodological contribution of this book is in its use of primary source material, specifically the writings of nineteenth century philosophers of science, as sources for constructing a robust description of the conventions for arguing with mathematics during this period. By relying on primary source material, it avoids problems of reliability and contextual sensitivity. In addition, it broadens the scope of materials available to rhetoricians for analyzing scientific argument and offers a means by which the division between common and special lines of argument can be made. These divisions are formulated positively based on examinations of the actual conventions articulated by a scientific discourse community rather than negatively as anything not existing in a particular treatise or set of treatises on rhetoric. Investigating the articulated conventions of science in conjunction with scientific arguments provides a broader and more accurate picture of what constitutes or does not constitute a common or special line of argument, and thereby what aspects of scientific argument are or are not being employed rhetorically.

Close Textual Analysis

While the historical analyses in the book are aimed primarily at establishing the conventions for mathematical and scientific argument, close textual analyses of the works of featured arguers offer insight into their specific choices of language, organization, and argument. These choices illuminate not only the persuasive goals and strategies of arguers, but also what these arguers may have believed about their audiences.

This type of analysis is conducted in this investigation using a number of pre-existing analytical categories in modern and classical rhetoric, such as *ethos, stases, loci,* value hierarchies, etc. as well as a detailed assessment of choices in language, organization, and argument in the text. The applications of these analytical categories are intended—in addition to their utilitarian function of illuminating the character and structure of the argument—to illustrate that categories for analyzing discourse and argument exist within rhetoric that might be profitably used to expand our understanding of the role of mathematics in scientific argument.

Audience Response Analysis

Finally, unlike the two previous analytical methods, which are primarily designed to illuminate argument conventions and strategies, the third method, audience response analysis, is designed to provide insight into persuasion. As some rhetoricians have pointed out, rhetorical analysts have made bold pronouncements about the persuasive affects of texts without supplying evidence from the audience to support their contentions.[4] The analysis in this book endeavors to make claims about the reasons that late nineteenth and early twentieth century biological researchers judged mathematical concepts and formulae to be reliable or unreliable grounds for arguments about variation, evolution, and heredity. As a consequence, it is necessary not only to discuss the conventions of scientific argument, but also the specific reasons given by audiences for accepting or rejecting them.

Examining the responses of individual audience members in conjunction with the conventions of scientific argument has a number of benefits. First, by examining the two together, it is possible to know whether members of a particular audience were or were not appealing to convention to support their praise or excoriation of a scientific argument. This knowledge provides a method of checking whether scientific conventions were taken seriously, considered unreasonable ideals, or not considered applicable in particular situations. Second, by examining individual responses it is possible to understand whether sources of good reasons for accepting or rejecting mathematical argument were limited to the conventions outlined in scientific philosophies. If alternative good reasons exist, their presence suggests that there may be values, beliefs, and truths from outside the confines of a specific scientific discourse community impacting the development of scientific knowledge. Their existence would indicate that broader, rhetorical lines of argument are implicated in reasoning about the validity of mathematical warrants in making scientific arguments about biological phenomena.

To make claims about what audiences might have thought about a particular application of a mathematical argument to some aspect of variation, evolution, and heredity, each chapter includes close readings of audience responses to the texts of the featured arguers. These responses are primarily from book reviews written at or around the time a featured text was released or, in the cases of journal articles, private or public responses offering commentary on the research. All audience responses are assessed by close textual analyses, singling out the reasons given by respondents for supporting or challenging the featured arguments. In addition, the reasons are compared to the featured arguer's theory of his audience to draw conclusions about why they may have succeeded or failed in their persuasive endeavor.

Preview of Chapters

Each chapter in the book is dedicated to investigating some aspect of the relationship between science, mathematics, and argument. In the first three chapters, the focus is on articulating the general conventions and limitations of mathematical argument in science. The remaining three chapters explore the rhetorical dimensions of making mathematical arguments in science.

Chapter 2, "A Proper Science," explores nineteenth century epistemological and ontological commitments about the appropriate relationship between mathematics and science by examining the philosophies of two of the period's most influential, natural philosophers. A close reading of William Whewell's, *Philosophy of the Inductive Sciences* (1840), and John Herschel's, *Preliminary Discourse on Natural Philosophy* (1831), suggests that quantification and mathematical reasoning were considered assets in the production of knowledge about nature because they contributed precision and rigor to scientific research and reasoning. Investigations of these texts also reveal the stages in which quantification and mathematical reasoning contributed to science and the process by which mathematical arguments might be elevated from hypothetical analogies to deductive laws of nature. The views of these two, influential philosophers about the status of mathematical arguments in science provides a framework for assessing the status of mathematical arguments, the choices of arguers as they seek to defend them, and the reactions of nineteenth century audiences as they move to accept or reject mathematical arguments as a legitimate means for making knowledge about nature.

The third chapter, "A New Species of Argument," examines the rise of the mathematical treatment of variation and evolution in Charles Darwin's work, *The Origin of the Species* (1859). It makes the case, contrary to most current scholarship on Darwin, that his work relies on common mathematical warrants—generally accepted lines of reasoning for making mathematical arguments in science—both to support and invent arguments about dynamic variation, relation by descent, and the principle of divergence of character. In addition, the chapter suggests that his use of mathematics for support and invention may have been a rhetorical move on Darwin's part to establish an ethos of precision and rigor for what he knew would be controversial positions. An examination of popular philosophies of science, which had been read by Darwin and were revered by many nineteenth century philosophers, suggests that the naturalist had attempted to follow the conventions for developing robust scientific arguments through the use of quantification and mathematical operations.

Whereas Chapter 3 investigates the use of common mathematical warrants as a means of enhancing the credibility of an argument, the fourth chapter, "Hidden Value," explores Gregor Mendel's reliance on special mathematical warrants—mathematical principles and formulae previously unsanctioned for argument about a particular subject—in his attempts to persuade his contemporaries to accept his theory of uniform particulate inheritance. Careful scrutiny of Mendel's arguments in "Experiments in Plant Hybridization" (1865) suggest that the mathematical principles of probability and combinatorics—mathematics "relating to the arrangement of, operation on, and selection of discrete mathematical elements belonging to finite sets or making up geometric configurations"—rather than being resources for description or verification of his experimental results, act as sources of invention for his experiments ("Combinatorial" Def. 2.). As a consequence, the mathematics function rhetorically as a creative analogy for imagining and arguing about nature before reasonable certainty had been established about its applicability to the case. I will argue that Mendel's confidence that other members of his audience would embrace the analogy as more than creative conjecture plays a central role in his failure to persuade them of the validity of his hereditary theory.

In addition to examining the analogical characteristics of mathematics in Mendel's argument, the chapter also investigates the complex network of ontological commitments which may have affected the reception of his mathematical arguments. A comparison of Mendel's ontological commitments with those of his contemporaries suggests that the principles on which Mendel founded his theory were, in many cases, directly at odds with those of mainstream studies of hybridization. As a consequence, the mathematical arguments, despite their analytical rigor, could be challenged on a number of grounds unrelated to their technical execution or their verification by experiment. This vulnerability suggests that Mendel's mathematical warrants existed in a competitive hierarchy of truths and values, some of which were either singly or collectively more compelling to his audience than the mathematical proofs which Mendel held in high esteem.

While Chapters 2, 3, and 4 probe the general divide between what was considered conventional or unconventional in mathematical arguments for mid-nineteenth century Continental and Victorian biological researchers, Chapters 5, 6, and 7 examine cases wherein explicit efforts are made to argue for the acceptance of special mathematical warrants as reliable descriptors of nature. These efforts reveal not only the reasons for the success or failure of particular attempts, but also further illuminate the scope of issues thought pertinent to, and reasons believed legitimate for, accepting or rejecting mathematical argument in science.

Chapter 5, "Probable Cause," investigates the rhetorical success of Charles Darwin's cousin, Francis Galton, in his endeavors to promote an analogy between the probabilistic law of error (as embodied in the bell curve) and the distribution of hereditary outcomes in human populations. Evidence from Galton's campaign to promote this analogy in his groundbreaking book, *Natural Inheritance* (1889), suggests that analogies between mathematics and nature had to be argued for, and that non-analytic rhetorical strategies played a substantive role in securing their acceptance.

To make this case, the fifth chapter investigates the rhetorical strategies in *Natural Inheritance,* the context of its publication, and its reception. An examination of the context of publication reveals that, like Mendel. the mathematical arguments Galton relied on to establish his conclusions about variation, evolution, and heredity were considered special warrants by his audience. A close textual analysis of *Natural*

Inheritance reveals that Galton, unlike Mendel, was aware of the lack of common acceptance of his warrants and looked for argument strategies and good reasons outside of the confines of the specialist discourse community of biological researchers to defend his conclusions. The use of general argument strategies such as narrative, visual argument, synonymia, and appeals to the values of his English Victorian readers are identified in his text, and evidence of their efficacy is presented from reviews of Galton's work.

Whereas Chapter 5 examines a successful effort to promote a novel mathematical approach amongst conventional biological audiences, Chapter 6 investigates the failure to expand it. "Behind the Curve" explores the mathematician Karl Pearson's efforts to develop—based on the success of Galton's arguments in *Natural Inheritance*—a purely mathematical model of inheritance based on the principle of probability. This exploration reveals that Pearson's dogged insistence that mathematics, and mathematics alone, was the key to understanding variation and evolution in natural populations, alienated his audience despite their general sympathy for Galton's analogy between the mathematical law of error and heredity. This failure suggests that even though the suitability of the analogy had been accepted, when mathematics was advanced as a *value* for doing science it could be challenged rhetorically.

The final chapter, "Weightless Elephants on Frictionless Surfaces," explores the early, twentieth century statistician R.A. Fisher's efforts to revitalize mathematical biology in the wake of Pearson and biometricians' failed attempts to persuade conventional biologists to side with their quantitative vision of variation, evolution, and heredity over Mendel's. This chapter concludes that Fisher believed he needed to construct and maintain a credible ethos for his own work as well as for the general program of mathematical argument in science to re-establish biometry as a viable approach to generating new knowledge about natural selection and evolution. An investigation of his early papers and seminal book, *The Genetical Theory of Natural Selection* (1929), suggests that by making his complex mathematical arguments accessible to scientists with limited mathematical training, and by arguing that mathematical arguments had the virtues of practicality and inductivity, Fisher made important strides in overcoming some of the final obstacles in the long and difficult road towards the synthesis of Mendelian genetics and Darwinian natural selection that were required for the emergence of population genetics.[5]

A Rhetorical Approach to Mathematics, Argument, and Science

By exploring the complexity of arguing mathematically in the study of variation, evolution, and heredity from the middle of the nineteenth to the beginning of the twentieth century, this book hopes to contribute to the understanding of mathematical argument in science and its rhetorical dimensions. What it reveals about mathematics in science is that its status as a warrant for making scientific arguments is not always secure, and in some cases, requires conventional and unconventional support to be accepted as legitimate. It also advances the possibility that mathematical descriptions and arguments in-and-of-themselves may not be sufficient reasons to accept a particular scientific conclusion. Instead, mathematics exists as one node in a complex hierarchy of good reasons in competition with other values, beliefs, and truths. These conclusions illustrate the rhetorical dimensions of mathematical argument, and should thereby further encourage rhetorical investigation into mathematics not only in science, but also in other areas, such as public policy, politics, and even theoretical mathematics.

Finally, this book is dedicated to showing how a rhetorical approach to argument analysis might contribute to the efforts of historians, philosophers, and sociologists of science in their quest to understand scientific knowledge. By carefully attending to the language, organization, and argument of specific texts, and the interrelations between texts, arguers, audiences, and contexts, rhetoricians offer methods for providing concrete textual evidence to support robust characterizations of the process of argument and knowledge-making in science that are contextually sensitive and empirically grounded.

2 A Proper Science: Mathematics, Experience, and Argument in Nineteenth-Century Science

> *Philosophy is written in this grand book—I mean the universe—which stands continually open to our gaze, but it cannot be understood unless one first learns to comprehend the language and interpret the characters in which it is written. It is written in the language of mathematics, and its characters are triangles, circles, and other geometrical figures, without which it is humanly impossible to understand a single word of it; without these, one is wandering about in a dark labyrinth.*
>
> —Galileo [1]

In the last twenty years, two major trends have emerged in the analysis of the rhetorical features of science, one taxonomic and another constructivist. In taxonomic approaches to analysis, scientific argument is identified and its effects explained using traditional concepts and terms from canonical treatises of rhetoric. This particular approach is employed in well-known rhetoric of science investigations, such as Lawrence Prelli's *Rhetoric of Science: Inventing Scientific Discourse*, and Jeanne Fahnestock's *Rhetorical Figures in Science.* While taxonomic approaches rely on established catalogues of *topoi*, tropes, and figures for analysis, constructivist investigations of argument, such as those undertaken by genre and action network theorists, focus on the features of communication and argument as they emerge and change in response to shifting needs within communities. In their analyses of scientific communication and argument, scholars like Carol Berkenkotter, Thomas Huckin, and John Swales, examine the conventions for scientific communication as well as how social, cultural, and institutional circumstances actively shape them (*Genre Knowledge*; *Genre Analysis*).

The analytical approach used in this book combines both of these theoretical perspectives and contributes a new set of resources for argument analysis. Throughout the text, taxonomic methods are employed to describe and explain the strategies for arguments used by researchers attempting to advance mathematical approaches to the study of variation, evolution, and heredity. For example, Chapter 3 examines Darwin's use of the commonplace "the more and the less" to make his case for dynamic variation in species.

Whereas the taxonomic aspect of this analysis provides language and a conceptual framework for describing argument, its constructivist dimension seeks real-world evidence of the conventions for arguing with mathematics in science. To understand this facet of argument, I turn to philosophies/methodologies of science—which have not, to my knowledge, been exploited as analytical resources—to understand conventions for arguing mathematically in science. To provide a context for the discussion in later chapters, this chapter examines in detail the two, nineteenth-century works on the philosophy and methodology of science: John Herschel's *Preliminary Discourse on the Study of Natural Philosophy* (1831) and William Whewell's *Philosophy of the Inductive Sciences* (1840). The conventions of mathematical argument described in these works provide a context for assessing not only the legitimacy of strategies used by arguers to advance mathematical approaches to variation, evolution, and heredity, but also the reasons for the success and failure of those strategies with nineteenth century, scientific audiences.

John Herschel and William Whewell

Scientific philosophies provide a valuable resource for understanding the choices scientists make when they argue. The influence of such philosophical texts was particularly strong in the nineteenth century, a period when the philosophy of science was not divorced from its practice. Philosophies of natural science were written by influential educators and practitioners who included in their works the latest information about the methods and state of knowledge in a broad range of scientific fields. As a result, they were not only read as theoretical documents, but also as handbooks describing the state of the discipline and the practice of science.

John Herschel (1792–1871) and William Whewell (1794–1866) were two of the nineteenth century's most eminently qualified writers on natural philosophy. They were both actively engaged in scientific research and publication, and both vigorously participated in developing important institutions of British science.[2] John Herschel is, and was regarded, as one of the great figures of Victorian science not only because of his tireless efforts in discovering and cataloguing astronomical phenomena, but also because of his ability to write lucidly about the finest points of scientific philosophy (Partridge xii-xiv). His skills of adaptation are exemplified in *Preliminary Discourse on the Study of Natural Philosophy* (1830), in which he presents readers with a thorough introduction to scientific philosophy and a clear explanation of scientific method.

The significant influence that the book had on Victorian science is evidenced not only by the fact that the text went through twelve editions, but also by the quality of the Victorian thinkers who vouched for its importance in the development of their own scientific thought. John Stuart Mill, for example, used it as the basis of his own work on scientific theory in his *System of Logic* (1843) (Partridge *xiv;* Canon, "John Herschel" 220–221). It also influenced James Clerk Maxwell's work in the *Discourse on Molecules*, and William Whewell's *Philosophy* (Canon, "John Herschel" 220; Kemsley).

Although not as eminent a producer of scientific knowledge as Herschel, William Whewell had a significant impact on mathematical and scientific education and natural philosophy in Britain. As an influential member of the faculty and administration of Cambridge from 1828–1866, Whewell pushed for the introduction of analytical mathematics at Cambridge, a move which brought the archaic mathematics curriculum at Cambridge up to date with Continental mathematical practices. He also supported the creation of a new Tripos for the natural sciences, which allowed students to focus their attention on the study of nature by relieving them of the extraordinary burden of having to have expert knowledge of mathematics to obtain honors in their studies (Herivel xiii).[3]

In addition to being in the avant-garde of institutional reform, Whewell was also a leader in nineteenth century discussions about the history, philosophy, and methodology of science and its relationship to mathematics. In 1837 he published *History of the Inductive Sciences*, which was "aimed at being, not merely a narration of the facts in the history of science, but a basis for the philosophy of science" (viii). In the text, Whewell traces the history of natural philosophy from the Greeks through the Middle-Ages and into the nineteenth century, critiquing the shortcomings and praising the advancements in scientific thought as it developed.

Whereas *History of the Inductive Sciences* explored and assessed the empirical development of a proper system for obtaining scientific knowledge, Whewell elaborates the epistemological characteristics of that system in *Philosophy of the Inductive Sciences* (1840). In part one, "Of Ideas," he offers an expansive discussion of scientific epistemology that addresses the relationship of thought and experience to the production of credible scientific knowledge. In part two, "Of Knowledge," Whewell describes the process by which he believed scientific knowledge was constructed and the specific methods by which knowledge of nature is obtained.

Although the philosophical positions in Whewell's *History* and *Philosophy* proved to be more controversial than Herschel's *Discourse*, they were nonetheless seriously regarded as important scholarship on scientific history, philosophy, and method by nineteenth century scientists. Both *History* and *Philosophy* ran three editions, and contributed to a lively debate about the foundations of scientific knowledge, which engaged important nineteenth century figures such as John Stuart Mill, Charles Darwin, and John Herschel.[4]

Because of their substantial influence on Victorian science and their attention to the role of mathematics in making scientific argument, the philosophies of science by Whewell and Herschel can be considered credible guides for understanding the challenges and benefits of making scientific arguments with mathematics. In addition, they represent opinions on two sides of an important philosophical division in the nineteenth century: between nativism, whose adherents believed that the source of knowledge about nature resides in the mind of the scientist; and empiricism, whose supporters suppose that the truth of nature inhered in nature itself, and could only be uncovered through experience and experimentation (Richards 2–3). Because they represented opposing sides of this debate, Whewell, nativism, and Herschel, empiricism, a combined analysis of their work affords a comprehensive view of the spectrum of opinion on correct procedure in Victorian science as well as common ground on the role of mathematics in making scientific arguments.

Rationality and Reality: The Stakes for Mathematical Argument

Whewell and Herschel took two divergent positions on the question of how reliable knowledge about nature could be attained. For Herschel the truths of nature existed in nature itself. Though the human mind was necessary for decoding the communications of nature, he did not consider the mind the source of true knowledge about nature. Instead, the ultimate source of natural knowledge was experience:

> We have thus pointed out to us, as the great, and indeed only ultimate source of our knowledge of nature and its laws EXPERIENCE, by which we mean not the experience of one man only, or of one generation, but the accumulated experience of all mankind in all ages, registered in books or recorded by tradition. (*Discourse* 76)

Whereas the collective, communal experience of nature was the wellspring of understanding for Herschel, Whewell located the universal principles of nature in the human mind. For Whewell the physical world as we perceive it presents us with data but not with the principles to comprehend the underlying relationships between phenomena. These principles could only be supplied by the mind. As a consequence, the mind and its faculties became the ultimate source of natural knowledge. The goal of science, therefore, was to uncover and clarify the vast and hidden laws of nature in the mind by observing and comparing data:

> In order to obtain our inference, we travel beyond the cases which we have before us; we consider them as mere exemplifications of some ideal case in which the relations are complete and intelligible. We take a standard and measure the facts by it; and this standard is constructed by us, not offered by Nature. (*Philosophy* 1: 49)

Though Herschel and Whewell supported two different positions on the ultimate source of knowledge about nature, they agreed that both experience and cognition were necessary complements in the construction of scientific knowledge. Proper science was the balance between the two. On the one hand, experience of natural phenomena was required because, without it, the products of reason, no matter how rationally rigorous, were simply elaborate fictions without purchase in nature. On the other hand, without the higher power of human reason, the hidden relationships between natural phenomena would be eternally locked away from view.

Questions about the appropriateness of mathematical argument and its benefit to the development of natural knowledge are caught up in this debate. Mathematics resided naturally on the mind/reason side of the Cartesian mind/body, reason/experience duality. This point is conceded by both Herschel and Whewell, and epitomized by Herschel in *A Preliminary Discourse on the Study of Natural Philosophy* when he writes:

> Abstract [mathematical] science is independent of a system of nature—of a creation—of everything, in short, except memory, thought, and reason. Its objects are, first, those primary existences and relations which we cannot even conceive not to *be*, such as space, time, number, order, &c. (18) [5]

Despite their position outside of nature, however, mathematical principles, operations, and symbols still had value in its characterization because it was with these conceptual tools that the invisible relationships between physical phenomena could be discovered. Herschel makes this point in the previous passage when he explains that relations of phenomena that have purchase in nature space, time, number, etc. can be conceived of in the abstract science of mathematics. Whewell makes the same point when he writes:

> All objects in the world which can be made the subjects of our contemplation are subordinate to the conditions of Space, Time, and Number; and on this account, the doctrines of pure mathematics have most numerous and extensive applications in every department of our investigations of nature. (*Philosophy* 1: 153)

Just as Herschel and Whewell agree that mathematical reasoning has a place in the interpretation of natural phenomena, both also agree that it only has validity if it is based on evidence from experience of nature.

Herschel recognizes the necessity of experience to mathematical reasoning when he writes,

> A clever man, shut up alone and allowed unlimited time, might reason out for himself all the truths of mathematics. . . . But he could never tell, by any effort of reasoning, what would become of a lump of sugar if immersed in water, or what impression would be produced on his eye by mixing the colors yellow and blue. (76)

Despite his opinion that the mind was the ultimate source of natural knowledge, Whewell also acknowledges the limitations of mathematics without experience. In an eloquent passage in volume one of *Philosophy of the Inductive Sciences,* he makes the point that without experience, mathematical knowledge of nature is impossible, and without mathematics, understanding the changes in natural phenomena is inconceivable.

> If there were not such external things as the sun and the moon I could not have any knowledge of the progress of time as marked by them. And however regular were the motions of the sun and moon, if I could not count their appearances and combine their changes into a cycle, or if I could not understand this when done by other men, I could not know anything about a year or month. (*Philosophy* 1: 18)

Though Herschel and Whewell emphasize different sides of the Cartesian split, both agree that experience and reason are necessary components of scientific knowledge. Reason—mental operations which reveal the hidden relationships between phenomena—opened the door for the participation of mathematical argument in the development of natural knowledge. However, experience—the data from observation and experiment—always acts as a limiting and shaping force on its contribution. These shared beliefs about the necessary balance between reason and experience represent the fundamental principles guiding Whewell's and Herschel's opinions about the possible strengths and potential weaknesses of mathematical argument as well as the manner in which robust, mathematical arguments about nature could be developed.

Mathematical Arguments and the Inductive Process

The delicate balance between experience and rationality plays itself out vividly in nineteenth century characterizations of "induction." In the process of induction, mathematics contributes the appropriate form for scientific arguments, while observation and experimentation provide the necessary content to verify the form. The constant check and balance between experience and reason is a key influence on the inductive process, dictating not only the steps by which mathematical argument might gain credibility, but also what arguers and audiences perceive to be the strengths and weaknesses of mathematical arguments.

For Herschel and Whewell, induction involved two distinct activities: the determination of causes, and the description of effects. Though both were interrelated, the development and use of mathematical arguments was directly implicated in the latter activity while only tangentially important to the former. As a consequence, their discussions of mathematical argument focus primarily on efforts to describe effects and their relationships to one another.

In combination, Whewell and Herschel identify four steps in the quantitative inductive process: quantification, formulization, verification, and extrapolation.[6] By examining these steps in detail, it is possible to understand how Herschel, Whewell, and presumably other natural researchers, perceived the possible strengths and potential weaknesses of mathematical argument, and how they could be raised from hypothetical to authoritative statements about nature.

Step One: Quantification

Both Herschel and Whewell are adamant that, without quantification, knowledge could not be considered "scientific." Whewell writes, for example, "We cannot obtain any sciential truths respecting the comparison of sensible qualities, till we have discovered measures and scales of the qualities which we have to consider" (*Philosophy* 1: 321). Herschel argues that, without quantification, argument could not be scientific because it would not have the necessary level of precision. Because human senses are not always sufficient to make the distinctions necessary to discover or describe changes in small or large-scale phenomena, the natural philosopher had to depend on precise quantification to establish reliable knowledge about nature. Herschel elaborates this point when he writes:

> In all cases that admit of numeration or measurement, it is of the utmost consequence to obtain precise numerical statements, whether in the measure of time, space, or quantity of any kind. To omit this, is, in the first place, to expose ourselves to illusions of sense which may lead to the grossest errors. (122)

Without precise quantitative data, Herschel explains, a scientific argument can never be considered reliable: "But it is not merely in preserving us from exaggerated impressions that numerical precision is desirable. It is the very soul of science; and its attainment affords the only criterion, or at least the best, of the truth of theories, and the correctness of experiments" (122).

If we accept the proposition that Herschel's and Whewell's opinions about the importance of quantification to the foundation of credible, scientific argument reflects and/or has influence on the opinions of other Victorian natural researchers, then we can assume that researchers making arguments about natural phenomena would aspire to use precise, quantified data to make their arguments compelling for their audiences. We can also assume that audiences assessing scientific arguments might praise or criticize them based on whether or not they were made using precise quantified data.

Step Two: Formulization

Once a standard for measurement is established and quantitative data is collected, the next step in quantitative induction is to describe the relationship in the data using a mathematical formula. With this step, the researcher proposes an analogy, wherein a particular relationship described by or derived from existing mathematical axioms is hypothesized to be analogous to the change in the natural phenomenon observed.

In the *Philosophy of the Inductive Sciences*, Whewell writes in detail about this process, suggesting that it has three steps: selection of the independent variable, construction of the formula, and determination of the coefficients (1: 382). He provides an example of the process using a hypothetical case in which astronomers attempt to discover the quantitative law describing how a particular star's position changes in the heavens. In the scenario, the researcher begins with observational data on the star, which shows that, after three successive years, the star has moved by 3, 8, and 15 minutes from its original place.

After consulting the existing data, he casts about for the appropriate category of change, or *Idea* under which a law might be constructed to describe the star's change in position.[7] If the investigation is to be quantitative, the *Idea* must come from one of four possible categories: space, time, number, or resemblance.[8] The researchers following the star settle on "time," which becomes the independent variable (*t*) for the formula with which they will express their law describing the star's movement.

After selecting an appropriate category of change, the scientist's next duty is to determine exactly how the measured phenomenon changes with respect to that category. If the category selected is "time," the researcher would ask, "How does the star's position change with respect to time?"; "Are the changes in time and position uniform? Are they linear? Are they cyclical?"

The change in the star's location of 3, 8, 15 minutes suggests that the alteration of its position with respect to the change of time is not regular. With the aid of his mathematical training, the researcher would quickly recognize that the series "can be obtained by means of two terms, one of which is proportional to time, and the other to the square of the time . . . expressed by the formula $at + btt$" (*Philosophy* 2: 383).

Once the apparent manner of change has been described in the formula, the magnitude of the coefficients—the fixed numerical constants by which the independent variables are multiplied—needs to be established. In the formula, $at + btt$, a and b are the coefficients. As Whewell explains, the magnitude of a and b could be established by figuring out what values were required to get the results described in the observations. To generate the series 3, 8, 15 from the equation $at + btt$ if time increases 1, 2, 3, etc., a must equal 2 and b must equal 1.[9]

For Whewell, the creation of a formula, which at this stage was considered a hypothetical representation of the change in a particular phenomenon, was an attempt to colligate (or collect) the instances of change under a single mathematical description. This move can be understood as an effort to make the case *for* a particular analogy between experience and reason (i.e., between observed data and known mathematical principles).

Analogy can be defined broadly as an argument *for* or *from* the resemblance between dissimilar constituents.[10] For example, Benjamin Franklin argued *for* accepting the resemblance between electricity and fluids in his efforts to explain the operation of the Leyden jar. Once this analogy was accepted, researchers such as Henry Cavendish used it as a basis *from* which to develop mathematical and mechanical explanations about the behavior of electricity (Jungnickel and McCormick 174–81). In *the New Rhetoric*, theorists Chiam Perelman and Lucie Olbrechts-Tyteca explain that analogies have two constituent parts, the *phoros* and the *theme* (373). The *phoros* is the part of the analogy with which the audience is familiar. It provides a structure, value, and/or meaning by which the unknown or unvalued *theme* can be understood or characterized. For example, in the analogy from Aristotle, "For as the eyes of bats are to the blaze of day, so is the reason in our soul to the things which are by nature most evident of all," the "eyes of the bat" and the "blaze of day" are the *phoros* because they represent a concrete relationship between knowable entities that the writer supposes the reader understands (*Metaphysics*, II: 933b, 10–11).[11] This concrete relationship is used to guide the reader in comprehending the abstract relationship in the *theme* between the "reason in the soul" and "things which are by nature most evident of all" (Perelman Olbrechts-Tyteca 373).

Based on Whewell's description of the process, the creation of a formula to express a particular change in a phenomenon can be construed as an argument *for* an analogy between reason and experience. The *phoros*—the suggested description of the change, warranted by the well-known axioms of mathematics—comes from the domain of reason. The *theme*—the perceived but vaguely understood change in the natural phenomenon[12]—is derived from a domain of experience that has been "translated" into a quantitative description to permit comparison. The final formula is the epitomized analogy, the proposed conclusion that the mathematical arrangement is a legitimate descriptor of a change, or the relationship between changes in a group of phenomena.

The benefit of making an analogy between quantified observations of nature and a mathematical formula whose components are related by strictly defined operations is that the result allows experience to be cast into a form that could be reasoned about clearly and rigorously. Because mathematical argument was governed by the established principles of logic at this time, conclusions reached through its use were considered credible if supported by sufficient evidence. Once verified, these conclusions could be used as axioms for making deductive arguments. In *A Preliminary Discourse on the Study of Natural Philosophy*, Herschel recognizes the rigor that mathematical form brings to arguments about nature:

> Acquaintance with abstract [mathematical] science may be regarded as highly desirable in general education, if not indispensably necessary, to impress on us the distinction between strict and vague reasoning, to show us what demonstration really *is,* and to give us thereby a full and intimate sense of the nature and strength of the evidence on which our knowledge of the actual system of nature, and the laws of natural phenomena, rests. (22)

According to Herschel's admonition here, for an argument to be considered sufficiently robust to be "scientific," it had to be made mathematically. This position reflects a consensus in nineteenth-century science that mathematical argument was the gold standard for making claims about natural phenomena. Given this sentiment, and the new self-consciousness engendered by works like Herschel's and Whewell's, there was a drive in all areas of natural investigation—even those in which there was no tradition of mathematization—to develop or use existing mathematical arguments to describe the changes and relationships between changes in natural phenomena (Cannon, *Science in Culture* 234–35).

Step Three: Verification

Once a formula is proposed, the next step in induction is to test the validity and limitations of the analogy by increasing the number of observations, and varying the conditions under which the data is gathered. This step is crucial when using mathematical arguments because it ensures that the necessary balance between the conceptual and the empirical is maintained.

The connection between the strength of conclusions and the number of trials/observations made to verify those conclusions was articulated at the beginning of the eighteenth century by Jakob Bernoulli in *Ars Conjectandi* (*The Art of Conjecture*) (1713). In the book, Bernoulli describes his famous "limit theorem," which states that the calculated *a posteriori* probability of an event (*p*) gets closer to the true *a priori* probability of an event (*P*) the greater the number of trials (*n*) that are conducted (Chatterjee 168).

Both Herschel and Whewell were generally acquainted with mathematical probability, as evidenced in their discussions of the method of curves as a way of identifying the "true value" in a set of observations (*Discourse* 130, 217–19; *Philosophy* 2: 398–400). They also seem to have been aware of Bernoulli's limit theorem for certifying the verity of the quantitative data and thereby the validity of the laws describing the relationship in the data. Whewell, for example, appeals to Bernoulli's principle when he writes: "In order to obtain very great accuracy, very large masses of observations are often employed by philosophers, the accuracy of the results increases with the multitude of observations" (*Philosophy* 2: 406).

While expanding the number of observations tests the verity of a mathematical analogy, increasing the variety of conditions under which trials are conducted helps determine its scope. Herschel advocates for both methods of verification, explaining that precise testing of quantitative hypotheses across a variety of circumstances can expose deviations in the data that might limit the analogy's scope or challenge its credibility:

> In the verification of a law whose expression is *quantitative,* not only must its generality be established by the trial of it in as various circumstances as possible, but every trial must be one of precise measurement. And in such cases the means taken for subjecting it to trial ought to be so devised as to repeat and multiply a great number of times any deviation (if any exists); so that, let it be ever so small, it shall at least become sensible. (*Discourse*168)

Whewell's and Herschel's discussion of the process of testing empirical laws reveals two obvious objections that might be brought against nineteenth-century researchers trying to establish conclusions using mathematical analogies. The objections of not doing a sufficient number of experiments, and not doing them under a sufficiently wide range of conditions, though not necessarily fatal to a particular argument, could force the arguer-scientist back into the field or laboratory to make further observations and experiments, or could require him to defend the breadth and depth of his empirical work. To support their claims, natural investigators could either remind readers about the scope or number of observations they undertook or limit their claims to the extent that they matched the level of proof their audience believed could be verified by the extent of the empirical evidence supplied.

Step Four: Extrapolation

Once a mathematical formula has been sufficiently tested to be considered a reliable analogy within a specific set of parameters, its argument status is changed. Instead of being the ends of the argument, it becomes the means. Herschel describes this transformation when he writes,

> These [empirical laws of nature], once discovered, place in our power the explanation of all particular facts, and become grounds of reasoning, independent of particular trial: thus playing the same part in natural philosophy that axioms do in geometry; containing . . . all that our reason has occasion to draw from experience to enable it to follow out the truths of physics by the mere application of logical argument. (*Discourse* 95)

The transformation from an argument *for* an analogy to an argument *from* an analogy is the result of the collapse of the *phoros* and the *theme*. This process is described by Perelman and Olbrechts-Tyteca, who write:

> Analogy finds a place in science, where it serves rather as a means of invention than as a means of proof. If the analogy is a fruitful one, theme and phoros are transformed into examples or illustrations of a more general law, and by their relation to this law there is a unification of the fields of the theme and the phoros. This unification of fields leads to the inclusion of the relation uniting the terms of the phoros and of the relation uniting the terms of the theme in a single category, and, with respect to this category, the two relations become interchangeable. There is no longer an asymmetry between theme and phoros. (396)

According to Perelman and Olbrechts-Tyteca, the process of validation, when successful, pushes the *phoros* and the *theme*, and the reason and experience, together to the point where any asymmetry between the two is lost. With this transformation, however, the question remains: "Is the formula still the epitome of an analogy?" Although Perelman and Olbrechts-Tyteca comprehensively describe the decomposition of analogy, they offer no comment on whether analogy, once it has gone through this process of decomposition, is still an analogy or something altogether different. If the hallmark of an analogy is an asymmetry between its *theme* and *phoros*, then empirical laws in which the *phoros* and *theme* are conflated seems to be something different altogether.

Once a mathematical formula has made the transition from an analogy to a law or principle of nature, it can be used as a warrant for making further arguments about phenomena both related and unrelated to the original subject of the induction. In "Of the Application of Inductive Truths," in *Philosophy of the Inductive Sciences,* Whewell offers astronomical tables as an example of how quantitative laws, once established inductively, are extended deductively to draw conclusions about subjects considered in the original induction, but not specifically used in the calculation of the laws: "*Tables* of great extent have been calculated, with immense labor, from each theory, showing the place which the theory assigned to the heavenly body at successive times; and thus, as it were, challenging nature to deny the truth of discovery" (2: 426). In Whewell's example, he cites the laws of planetary motion, arrived at by observing a few heavenly bodies, and then extrapolated in tables to describe the motion of other like objects, as instances where deduction is applied to the same class of subjects that were considered in the original induction.

In other cases, an empirical law can be used to predict phenomena which were not the original subjects of the induction by which the law was established. Herschel cites Newton's and others' applications of the theory of gravitational attraction to deduce the anomalies in the motions of the planets as an example where inductively established laws lead, via deductive extrapolation, to arguments about phenomena not considered under the original laws:

> We must set out by assuming this law [of gravitational attraction] . . . we then, for the first time, perceive a train of modifying circumstances which had not occurred to us when reasoning upwards from particulars to obtain the fundamental law; we perceive that *all the planets* attract *each other* . . .

> and as this was never contemplated in the inductive process. (*Discourse* 201)

By developing further mathematical calculations from the law of gravity to describe the amount of influence planets have on one another, and then using the results to predict eccentricities in their orbital paths, Newton and others seeking to verify or extend his theory of gravitation proved that the law of gravity accounted for anomalies in planetary motion that had previously puzzled researchers. If the theory of gravity had not been able to suitably account for these effects, for which it had been deduced to be the cause, then the credibility of the law would have been in jeopardy (202).

In the final stage of induction, mathematical analogies make an important transition from tentative conclusions to generally accepted warrants for further arguments. Although the laws established from these analogies are still open to emendation and clarification, they have passed an important threshold after which they are generally considered accepted principles of nature. As a result of their new status, they can serve as axioms from which extrapolations can be made about subjects that fall under their jurisdiction, or about phenomena not originally considered. In this capacity, mathematical warrants serve as engines of invention, suggesting new pathways for expanding natural investigation.

Conclusion

By examining in tandem the works of two of the most influential, nineteenth-century philosophers/methodologists of science, this chapter has endeavored to provide the background for assessing what constitutes the usual or commonly accepted criteria for making mathematical argument in science in the nineteenth century, and the appropriate stages by which mathematical warrants were thought to develop. Though fundamental disagreement existed between Herschel and Whewell on the ultimate source of natural knowledge, they both agreed that without quantitative laws, nature's intricate and sometimes impossible-to-observe operations could never be brought to light. They also believed that the strength of mathematical arguments resided in their capacity to illuminate these operations in a precise and rigorous manner, which spared natural researchers from the weakness of memory and the illusions of experience. These obvious benefits of mathematics helped it to persist in biological investigations of variation, evolution, and heredity despite general disagreements over the applicability of mathematical laws to biological phenomena. Conclusions supported by quantified data or mathematical operations could be considered more precise and rigorous than those that did not.

Despite obvious strengths, mathematical reasoning could be challenged on the grounds that it did not accurately reflect experience. As a consequence, mathematical formulae and reasoning had to be tested against evidence from repeated observations and experiments under a variety of conditions. As both modern and historical cases reveal, it is only in the presence of data that mathematical applications and arguments thrive in science. Chapter 4, for example, examines how Mendel's work fell into obscurity because it lacked a broader data set to support its conclusions, and Chapter 5 explores how Galton's work succeeded in part because of his herculean efforts to collect data in support of his theory of inheritance.

In addition to describing the qualities that make mathematical argument robust, this chapter has also illustrated the stages by which mathematical knowledge achieves legitimacy. Understanding where argument is perceived to be in this process provides insight into why a particular argument may or may not be considered rhetorical, and for what reasons. When I use the term "rhetorical" here, I am talking about argument which is probable rather than certain; argument which produces agreement from a variety of sources, which includes, but is not limited to, emotions, beliefs, and values; and finally, argument that relies on a number of general strategies/tools for argument, including figures, tropes, and *topoi* as means to secure agreement and establish understanding. Scientific arguments at the beginning stages of mathematization take on a rhetorical dimension because they rely heavily on the prestige accorded to mathematical deductive rigor and precision to make their scientific case, which initially has only a limited amount of inductive, empirical evidence to support it. In the middle stages of the process, the rhetorical dimension of mathematical arguments shifts from a reliance on the ethos of rigor and precision of mathematics to a dependence on analogy to establish understanding and secure agreement. Finally, in the last stages of quantitative induction, fused analogies are no longer rhetorical because they can be used as a common ground for further argument. However, because the possibility of a challenge always exists, they have the potential to lose their status as reliable warrants for scientific argument and fall once again into the realm of the probable, the rhetorical.

In the chapters that follow, the conventions for mathematical argument set out in Herschel's and Whewell's philosophies provide an epistemological framework for assessing the strategies of arguers as they attempt to advance mathematical programs for the study of variation, evolution, and heredity, and their successes or failures in making their cases. These investigations illustrate the utility of scientific philosophies and methodologies in understanding the epistemological context of mathematical argument in science, and it's possible rhetorical dimensions.

3 Evolution by the Numbers: Mathematical Arguments in *The Origin of Species*

I have deeply regretted that I did not proceed far enough at least to understand something of the great and leading principles of mathematics; for men thus endowed seem to have an extra sense.

—Charles Darwin

So every idea of Darwin—variation, natural selection, sexual selection, inheritance, prepotency, reversion—seems at once to fit itself to mathematical definition and to demand statistical analysis.

—Karl Pearson

Because Darwin's impact on the social and scientific developments in the nineteenth and twentieth century has been so large, his work, particularly *The Origin of Species*, has garnered a lot of attention from scholars in a variety of fields, including rhetoric, history, philosophy, and literature. In rhetoric alone, his persuasive efforts have been the focus of papers by authors such as: John Angus Campbell, Jeanne Fahnestock, Alan Gross, and Carolyn Miller (Campbell, "Perspective," "Polemical," "Rhetorician," "Invisible," "Believed;" Fahnestock, "Series Reasoning;" Gross, "Taxonomy;" and Miller and Halloran). Their works explore different aspects of his rhetorical strategy in The *Origin of the Species*, including the use of analogy between the human breeder and nature, to help his audience understand the operation of natural selection and the importance of the rhetorical figures *incre-*

mentum and *gradatio* in making the argument about variation and diversity among groups of organisms.

Despite the wide range of topics and issues in argument covered by rhetorical scholars, there has as yet been no substantive discussion about the rhetorical importance of mathematics in making arguments in *The Origin of the Species*. The purpose of this chapter is to offer arguments and analyses that suggest that Darwin relies heavily on mathematical elements such as quantification and basic arithmetical operations for support and invention of his arguments for dynamic variation, relation by descent, and the principle of divergence of character in *The Origin of Species*. It will also make the case that, by following the best practices of quantitative induction, Darwin hoped to establish an ethos of precision and rigor for his work which was commensurate with the rising importance of quantification to the study of biological phenomena in the middle of the nineteenth century.

Mathematical Darwin?

Though historians and philosophers of science have expended considerable effort tracing the development of different mathematical fields and examining their political, cultural, and even rhetorical influence (e.g., Cullen; Patriarca), they have not, with rare exceptions, taken up investigations into the role of mathematics in Darwin's arguments. A survey of eleven books and nine articles published by historians, philosophers, and rhetoricians of science, most published in the last twenty-five years, reveals that few texts associate Darwin's arguments with mathematical reasoning (Appendix A). Those texts that do associate the two predominantly comment either on the lack of mathematical reasoning in the text, or on Darwin's inability to use mathematics to make his case (Ghislen; Hull; Gale; Depew and Weber). Only four assign any real importance to mathematics in Darwin's arguments (Browne; Schweber; Parshall; Bowler).

Mathematics in *The Origin of Species*

The previous examination of selected books and articles in the history, philosophy, and rhetoric of science suggests that many modern scholars do not believe or have not considered mathematical argument as an important facet of Darwin's persuasive strategy in *The Origin of Species*. These results raise the question, "If mathematics plays such an important role in Darwin's argument, why is it that so few scholars in rhetoric and history bothered to write about it?"

A cursory review of the text itself reveals that there are very few places where mathematical symbols, numbers, tables, equations, etc. are used. This scarcity of mathematical notation is puzzling even to those who argue in favor of the importance of mathematics in *The Origin of the Species*, like historian Janet Browne, who comments on the scarcity of mathematics in the text:

> That Darwin's botanical arithmetic has been neglected by historians is partly his own fault. In *On the Origin of Species,* he barely referred to his botanical statistics or the long sequence of calculations which he had undertaken from 1854 to 1858. He compressed and simplified these into a few meager paragraphs, giving his reader only six pages of statistical data to fill out the discussion of "variation of nature" in Chapter II. (53)

Despite its absence in the actual text, a brief review of Darwin's notebooks, letters, the published manuscript of his "big species book," and *Variation of Plants and Animals under Domestication*, reveals the extent to which mathematics influenced the development of his theories.[1] In these publications, Darwin supplies his readers not only with lists of quantitative evidence and calculations, but also with occasional glimpses of the degree to which these data and calculations helped him formulate his conclusions.

The existence of precisely quantified data and calculations in these extrinsic sources, however, still does not explain why, if they were so important to Darwin's argument, the majority of them were left out of his text. The answer to this query is provided by Darwin himself in the introduction to *The Origin of the Species.*

> I can here give only the general conclusions at which I have arrived, with a few facts in illustration, but which, I hope, in most cases will suffice. No one can feel more sensible than I do of the necessity of hereafter publishing in detail all the facts, with references, on which my conclusions have been grounded; and I hope in a future work to do this. (4)

Here Darwin explains that he is able to give only a general outline of his theory, and as a result, has to forgo presenting all of the data and discussion that he might have otherwise provided. The reason for this brevity is that he has been rushed into publication by the emergence of Alfred Russel Wallace's theory of evolution which, for all intents and purposes, offered the same conclusions as his own. Additionally, Darwin's lack of specific, quantitative detail may have been a strategy to make his work accessible to a wider readership for whom a text dense with quantitative data and arithmetical calculations would have seemed too formidable (Beer viii).

Besides the infrequency of quantified data and operations in *The Origin of the Species*, influential historians discussing the arguments in the text, notably David Hull, have made the case that Darwin could never have integrated mathematical reasoning into his arguments because this type of reasoning was deductive and could not be brought into the service of an inductive theory. In his book, *Darwin and His Critics*, Hull takes the position that because Darwin was developing arguments in the non-physical sciences, deductive mathematical reasoning could not aid him in prosecuting his argument:

> Darwin could not help but know the crucial role which mathematics had played in physics, since Herschel had repeatedly emphasized it in his *Discourse,* but it did not seem to be in the least useful in his own work in biology. . . . For Darwin, mathematics consisted of deductive reasoning, and he distrusted greatly "deductive reasoning in the mixed sciences." In his own work, he seldom was presented with a situation in which he could use such deductive reasoning. He was constantly forced to deal in probabilities, and no one could tell him how to compute and combine such probabilities. (12–13)

Hull's assessment of the impossibility of Darwin's use of mathematics rests on the assumption that Darwin believed that mathematical reasoning was deductive, and therefore could not be used in inductive arguments. Like Hull, rhetorical theorists most likely also miss the rhetorical dimension of the text because they assume that mathematical warrants are deductive. Though this does not preclude them necessarily from functioning in Darwin's text, it does remove them, in the minds of most rhetorical analysts, from being the focus of a rhetorical investigation. To my knowledge there has been no explicit statement such as Hull's that this consideration has kept rhetoricians from examining the mathematical aspects of Darwin's arguments. However, this restriction is articulated in influential theoretical texts such as Philip

Davis and Rueben Hersh's "Rhetoric and Mathematics," in which they write that, in the minds of most rhetorical scholars, "If rhetoric is the art of persuasion, then mathematics seems to be its antithesis. This is believed, not because mathematics does not persuade, but rather that it seemingly needs no art to perform its persuasion" (53).

Philosophers, historians, and rhetoricians of science have not recognized an important role for mathematics in Darwin's arguments. However, careful examinations of early nineteenth century botany and geology, a detailed investigation of Darwin's ideas in his notebooks and letters, and a close textual analysis of the arguments in The *Origin of the Species*, reveal that quantification and basic mathematics were important to his work. They show that mathematics played a central role in Darwin's formulation and defense of his arguments, including his rhetorical efforts to establish an ethos of precision and rigor for his work.

Keeping Count: The Rise of Statistics in the Nineteenth Century

One of the fundamental characteristics of robust science in both modern and Victorian characterizations is quantification. Without the ability to "translate" natural phenomena into the language of numbers, induction leading to the formation of empirical laws could not commence. It was, therefore, the first step in the formation of any science to discover the method or system of measure on which quantitative induction could be founded.

Although various attempts had been made in the eighteenth century by Linnaeus and others to quantify certain aspects of biological research (such as the classification of leaves and reproductive organs in plants), they were all considered, at least by nineteenth century standards, artificial, and therefore not sufficient for the basis of a quantitative science. At the beginning of the nineteenth century, however, two important developments afforded new opportunities for advancing quantitative investigations of organic phenomena. The first was the increased interest in and use of statistics. The second was the discovery of fossils of extinct organisms, whose forms were completely alien from existing flora and fauna, which focused attention on questions about the origin and dispersion of organic forms. In his search of evidence and arguments for *The Origin of Species*, Darwin was influenced by both of these developments, which inspired him to cultivate quantitative evidence and mathematical arguments to support his theories of variation and evolution.

Though vital statistics (numbers of births and deaths) had been collected since the seventeenth century by religious and political organizations, the number of investigations and degree of attention to their results was limited to very small audiences.[2] At the beginning of the nineteenth century, however, social, political, and economic contingencies converged to create what statistical historian Harald Westergaard dubs the "The Era of Enthusiasm" for statistics (136), and Ian Hacking calls a period with "a professional lust for measurement" (5).

There is no single, agreed upon cause for this sudden interest in and collection of statistics. Some historians attribute it to the need to for precise measurement required by the Industrial Revolution, which gathered momentum at the beginning of the nineteenth century (Hacking 5). Others argue that it was the result of a sudden increase in the availability of statistical information that followed the end of the Napoleonic wars (Chatterjee 267). Yet others contend, perhaps most convincingly, that the supply of statistical data increased to meet a greater demand by governments who required quantitative data in order to make better informed political decisions and more persuasive policy arguments (Westergaard 141; Cullen 19–20; Patriarca 13–14). In particular, governments required statistics on birth and death rates as well as the resources of their domain and the domains of other nations to make rational economic policy decisions.

The statistical fever that had grabbed hold of politicians and moral philosophers in the early decades of the nineteenth century also infected geologists and botanists who were exploring the vast biodiversity of the Americas and Australia. Like political economists, they began in earnest to gather quantitative data on organic populations and the conditions under which they thrived. However, unlike their counterparts, the ends for their statistical efforts were affected by important scientific questions raised by new geological theories which assumed a dramatically older earth and grappled with new fossil evidence of organisms unlike any flora or fauna known to Victorian science. These discoveries, which challenged the tenants of the Christian doctrine of creation, encouraged investigations attempting to reconcile, to some extent, the scientific evidence with religious doctrine.

These efforts gave rise to a new field of biogeography, whose aim was to answer fundamental questions about organic populations, including: "What causes influence the thriving or extinction of particular species?"; "What is the distribution of species and genera upon the globe?"; "What is the population of any given species?"; "How can we account for the appearance of new species throughout geological time?"; and "What are the laws by which plants and animals of different parts of the earth differ?"[3]

Part of the spirit of this new field was that these questions needed to be answered not through classification of organisms and minerals, but rather through the juxtaposition of quantitative facts about climate, population, and location. Alexander von Humboldt, one of the early founders of biogeography, proclaims this goal in *Aspect of Nature in Different Lands and Different Climates* (1849):

> Terrestrial physics have their numerical element, as has the system of the universe, or celestial physics, and by the united labors of botanical travelers we may expect to arrive gradually at a true knowledge of the laws which determine the geographical and climactic distribution of vegetable forms. (108)

The path towards a new biogeographical physics was laid down in works such as Alphonse de Candolle's "Essai Elementaire de Geographie Botanique" (Elementary Essay on Botanical Geography) (1820), Robert Brown's *General Remarks, Geographic and Systemical, on the Botany of Terra Australis* (1814), and Joseph Hooker's *Botany of the Antarctic Voyage*, Vol. 2 (1853).[4] In their texts, statistics on temperature, elevation, size of organic populations, and the size and distribution of genera and species were used to make arguments bearing on the questions of distribution, variation, origination, etc. of plants and animals.

Robert Brown's work exemplifies the biogeographer's efforts to use basic, arithmetical operations and quantified data to compare and make arguments about variation in organic phenomena and the relationship of this variation to geographic and climactic conditions. In the *Botany of Terra Australis*, for example, Brown tests the assertion commonly held by nineteenth century botanists that dicotyledonous plants outnumber monocotyledonous plants by examining whether climate affects the numbers of either type in the general botanical population: [5]

> With a view to determine how far the relative propositions of these two classes [dicotyledons and monocotyledons] are influenced by climate, I have examined all the local catalogues or Floras which appear most to be depended on. . . . The general results of this examination are that from the equator to about 30° of latitude, in the northern hemisphere at least, the species of dicotyledonous plants are to monocotyledonous plants as about 5 to 1 . . . and that in the higher latitudes a gradual diminution of dicotyledonous takes place, until at about 60° N. lat. and 55° S. lat. they scarcely equal half their intratropical proportion. (*Miscellaneous Botanical Works* 8)

In the passage Brown draws on quantitative descriptions of location and previously tabulated statistics on the number of species of each sort of plant as well as calculated ratios to describe the limits of the geographical distribution of dicotyledonous and monocotyledonous plants. He concludes, based on the data and calculations, that dicotyledonous plants are abundant near the equator but become less abundant in northern latitudes. This conclusion provides precise quantitative detail supporting what was otherwise an anecdotal assumption about the difference in dicotyledonous and monocotyledonous plants in the general population of flora. It also supplies new information about the relationship between location and the thriving of dicotyledonous plants, which was previously unknown.

Darwin and the Biogeographers

The quantitative data and mathematical methods used by Brown and other biogeographers inspired Darwin to develop quantitative/mathematical arguments about the phenomena of variation and evolution. Evidence that Darwin was inspired by their methods can be found in his reading habits, in the private thoughts he recorded in his notebooks, and in his letters discussing the construction of his arguments for *The Origin of the Species.*

A brief assessment of Darwin's reading habits during the development of *The Origin of the Species* reveals that Darwin was familiar with the central works of biogeography and read them as he was developing his ideas for his masterwork. He was most likely introduced to them by his mentor John Henslow, who presented him, before he left Cambridge, with Humboldt's *Personal Narrative of Travels to the Equinoctial Regions of the New Continent during the Years 1799–1804*

(1814–1829) in which the explorer discusses the importance of quantification in the discovery of natural laws and offers statistics on temperature, altitude, longitude and latitude, etc. for various regions and flora in South and Central America (Schweber 205). It is clear that Darwin read this book and found it important because his copy contains extensive notes. According to the compiler of Darwin's marginalia, Mario A. Di Gregorio. Humboldt got Darwin "thinking about the distribution and the relation of organism to organism in the context of isolation, extinction, and the breeding of wild and domesticated animals" (xxxv). In addition to Humboldt, Darwin also owned multiple editions of Lyell's *Principles of Geology* (1st, 5th, 6th, 7th, 9th, 10th, and 11th) (Di Gregorio 530–44). In the fifth edition, volume three, chapter five, "Laws which Regulate the Distribution of the Species," discusses at length the important figures of biogeography and their work. Like Humboldt's *Personal Narrative,* Darwin marked and commented heavily on the text, providing evidence for his interest in their ideas (Di Gregorio 535–36).

When Darwin returned home and began developing the ideas that would be presented in *The Origin of the Species,* he continued to show interest in the work of biogeographers, as references to their works and ideas in his notebooks testify. At the end of *Notebook C,* in a section titled, "To be Read," for example, Darwin lists "Brown at end of Flinders & at end of Congo voyage, Decandolle. Philosophie, or Geographical Distribution. <<in Dict. Sciences. Nat. in Geolog Soc.>>" as texts which he believed would be beneficial to the development of his ideas (268).[6] Of these two works, Alphonse de Candolle's "Essai Elementaire de Geographie Botanique" was of particular importance to developing his ideas in *The Origin of the Species.* Of all the books listed in Gregorio's *Charles Darwin's Marginalia,* Darwin's copy of "Gèographie Botanique" has the most marginalia, prompting Di Gregorio to write that the work "seems to be a catalyst for much thinking around distribution, the struggle for existence, isolation, and consequently selection" (xxxiv).

Not surprisingly, the writings in the notebooks themselves echo the biogeographers' sentiments about the importance of developing a quantitative/mathematical approach to the study of plant and animal life. In *Notebook D,* for example, Darwin echoes Humbolt's sentiment that investigators of organic phenomena could and should emulate the quantitative inductive methods used in celestial and terrestrial physics:

> Astronomers might formerly have said that God ordered each planet to move in its particular destiny.—In the same manner God orders each animal created with certain form in certain country, but how much more simple, & sublime power let attraction act according to certain laws such are inevitable consequent let animal be created, then by fixed laws of generation, such will be their successors.—let the powers of transportal be such and so will be the form of one country to another.—let geological changes go at such a rate, so will be the numbers & distribution of the species!! (*Notebook B*, 101–102)

In this passage, Darwin reveals a clear link in his thinking between the methods of astronomy and the study of organic phenomena ascribed to by biogeographers. Just as astronomers create general laws by quantifying the period of the revolution of celestial bodies, their distance from one another, the amount of space they sweep out in a given period, etc., the biogeographer could arrive at quantitative laws describing the number and distribution of the species by determining the rate of generation of organisms, the rate in change of geological conditions, and the power of transport.

Darwin's reading habits and his notebooks supply evidence that he was considering the idea that natural laws might be discovered by working inductively from quantified evidence as he was gathering his thoughts for what would become *The Origin of Species*. However, it was not until the mid-1850s, when he began writing the manuscript of *Natural Selection* (his "big species book"), that there is evidence that Darwin began to put these ideas into practice by gathering statistical data and making arithmetical calculations to test his theory of evolution and build his arguments.

In 1855 Darwin began lengthy correspondences with Asa Gray and H. C. Watson in which he asked the botanists to supply him with information about genera and species, and discussed with them calculations of the ratios of varieties to genera in large and small genera. The quantified data and mathematical calculations discussed in these letters serve as sources of invention and argument for a *relationship of descent* and the *principle of divergence of character* in *The Origin of the Species*.

In his second letter to Asa Gray, on August 24, 1855, for example, Darwin hints to Gray that if he could get a reliable systematist to help him identify "close species," (i.e., species that closely resemble one another) he could calculate whether there was a propensity for larger genera to have more of these types of species than smaller genera:

> It occurred to me that if I could get some good systematists . . . to mark (without the object being known) the close species in a list; then if I counted the average number of the species in such genera, & compared it with the general average . . . of the species to the genera in the same country; it would, *to a certain extent,* tell whether on average the close species occurred in the larger genera. (Darwin to Gray, August 24, 1855)

Darwin makes a similar request to Watson, who obliges him by marking close species in his catalogues of English plants (Watson to Darwin August 17, 1855). Using Watson, Gray, and other botanists' marked compendiums of species, Darwin searched for proof of a general pattern in the ratios of closely related species in large and small genera. He hoped to find that larger genera had a greater number of close species, while smaller genera had fewer, more distinct species. He reasoned that if this pattern did exist it would support his theory of variation and relation by descent because larger, more successful genera would be producing new species that would be recent and closely related. Conversely, unsuccessful genera would not be producing newer species, and some of the older species they had produced would have died out, leaving gaps between the existing species and making them appear less closely related (Watson to Darwin November 19, 1854 n2). Evidence of this strategy appears in the draft of the big species book from which *The Origin of the Species* was abstracted (Table 1).

The Origin of Species: A General Overview of the Argument

An examination of Darwin's letters, notebooks, and reading lists reveals that he was familiar with ideas and methods of biogeography and that he adopted them in his search for evidence and arguments to support his theories of variation and evolution. In order to make the case that mathematics is an important aspect of Darwin's argument in *The Origin of the Species,* however, it is imperative to show that mathematical argumentation exists in the text and to understand what role it plays in supporting Darwin's conclusions. Towards this end, the next two sections examine the arguments in chapters two and

Table 1. Arithmetical Comparisons of the Ratios Between Species and Varieties in Large and Small Genera. Reprinted from Charles Darwin, *Charles Darwin's Natural Selection*, Ed. R.C. Stauffer, p. 149.

VARIATION UNDER NATURE

TABLE I[1]

For particulars on the works here tabulated and on the few corrections made, see the Supplement to this Chapter.	The numerators in the columns give the number of species presenting varieties; the denominators the number of species in the larger and smaller genera: these fractions are all reduced to common denominators of a thousand for comparison, and are printed in larger type to catch the eye. The right hand rows of figures in the three columns, with decimals, show the average number of varieties which each varying species has,—thus the number 1.50 shows that each two varying species have on average between them three varieties.

	Larger Genera	*Smaller Genera* (including those with single species)	*Genera with a single species*
Great Britain. Bentham			
Great Britain: Babington—Larger Genera with 5 species and upwards, smaller with 4 species and downwards [Pencil note by C.D.: Write this column larger'.]	$\frac{101}{663} = \frac{152}{1000}$ 1.40	$\frac{89}{745} = \frac{119}{1000}$ 1.30 [Pencil note by C.D.: 'Write this larger'.]	$\frac{24}{255} = \frac{94}{1000}$ 1.50
Great Britain, Henslow—Larger Genera with 5 species and upwards, smaller with 4 species and downwards. The Varieties are divided into two groups, the less strongly marked, and those which have been ranked by some eminent Botanists as species. Lesser Vars:	$\frac{69}{560} = \frac{123}{1000}$ 1.55	$\frac{67}{692} = \frac{96}{1000}$ 1.40	
Stronger Vars:	$\frac{33}{560} = \frac{58}{1000}$ 1.33	$\frac{29}{692} = \frac{41}{1000}$ 1.20	
Great Britain—London Catalogue (1853) (see Supplement for nature of Varieties)—Larger Genera with 5 species and upwards, smaller with 4 species and downwards	$\frac{97}{616} = \frac{157}{1000}$ 1.35	$\frac{85}{642} = \frac{132}{1000}$ 1.27	
Great Britain—London Catalogue—forms ranked as species in this catalogue but which have been thought by some authors to be varieties. In this second line, larger genera with 5 species and upwards, smaller with 4, 3, and 2 species	$\frac{57}{559} = \frac{101}{1000}$	$\frac{14}{377} = \frac{37}{1000}$	

1 [Darwin's holograph draft for this table is in ULC vol. 16.1, fol. 167.]

four of text as well as the peripheral documents associated with them. This investigation reveals that Darwin employs mathematical argument in the book, and that this mathematical argumentation plays an important role in helping him invent and/or support his arguments for the existence of a process of *dynamic variation* and the *principle of divergence of character.*

To understand the significance of the arguments in chapters two and four in the overall scheme of Darwin's argument, it is useful to review the primary conclusions he attempts to establish in the text. The basic arguments in are: (1) that organisms are highly plastic and can be made to vary to a great degree, (2) that variation accumulates over time, resulting in populations of organisms that were once related becoming physically distinct, (3) that the spread of variation in nature is the result of natural selection, and (4) that the more diversity in a species or genera the more likely it is that its members will successfully reproduce.

These primary arguments are introduced and developed in the first four chapters of the book. The other chapters of the text are concerned with presenting qualitative evidence from geology, animal behavior, comparative anatomy, and other areas of knowledge that support Darwin's four main arguments and his efforts to address possible disputations of his position.

Chapter II: Variation under Nature

In the second chapter of *The Origin of the Species*, "Variation under Nature," Darwin argues for the possibility of selection without human intervention through the process of natural selection. He achieves this goal with the help of quantified comparisons using arithmetical operations that prove that not only are taxonomic categories of species fuzzy, but also that this fuzziness can be accounted for by conceiving of diversity as the result of the dynamic process of continual variation, revealing a relationship of descent between the different levels of the taxonomic hierarchy.

In the first portion of chapter two, Darwin sets up his argument by refuting the position of special creationists who believed that each species identified in the taxonomic hierarchy marked a unique creation that was readily identifiable by the existence of an indelible set of features. He argues that if this position is correct, then there should be a definite consensus about which organisms belong in a particular category. In order to test the veracity of this assumption, Darwin quantitatively compares the categorization statistics made by experts in the field, including H.C. Watson:

> Compare the several floras of Great Britain, of France, or of the United States, drawn up by different botanists, and see what a surprising number of forms have been ranked by one botanist as a good species, and by another as mere varieties. Mr. H.C. Watson, to whom I lie under deep obligation for assistance of all kinds, has marked for me 182 British plants, which are generally considered as varieties, but which have all been ranked by botanists as species. . . . Under genera, including the most polymorphic forms, Mr. Babington gives 251 species, whereas Mr. Bentham gives only 112, - a difference of 139 doubtful forms! (41)

Using their own data, Darwin reveals that even experts in plant identification and categorization come to astonishingly little agreement about which organisms should be ranked as varieties and which as separate species. By casting doubt on the fixity of taxonomic categories, he creates an opportunity to present his own theories of evolution and natural selection, which he believes more adequately account for the data.

He opens the second portion of the chapter by clearly laying out his position:

> Hence I look at individual differences, though of small interest to the systematist, as of high importance for us, as being the first step towards such slight varieties. . . . And I look at varieties which are in any degree more distinct and permanent, as steps leading to more strongly marked and more permanent varieties, and at these latter, as leading to sub-species and to species. . . . I attribute the passage of a variety, from a state in which it differs very slightly from its parent to one in which it differs more, to the action of natural selection in accumulating . . . differences of structure in certain definite directions. (44).

In these lines, Darwin presents a vision of diversity in nature as a dynamic process rather than as a static condition. He argues that the small differences observed in individual organisms can spread by descent throughout successive generations, making the offspring of those individuals slightly different from the general population from which they originated. These differences can widen through the continued accumulation of variation and eventually transform distinct varieties into distinct species. This dynamic process, Darwin argues, can be attributed to natural selection, which he defines as, "the preservation of favorable variations and the rejection of injurious variations" (68).

Once he has established his position on the source and character of diversity, Darwin presents arguments to attempt to link the size and range of a group of organisms at a particular level of the taxonomic hierarchy to the number of subordinate categories of organisms associated with that group. In order to do this, he depends both on arithmetical calculations using quantitative data and on the rhetorical/logical commonplace (*topos*) of *the more and the less* with which he establishes a connection between size/range of a population and the characteristic of diversity. In Book II of the *Topics*, Aristotle explains this strategy of argument:[7]

> Moreover, argue from greater and less degrees. There are four commonplace rules. One is: see whether a greater degree of the predicate follows a greater degree of the subject. . . . For if an increase of the accident follows an increase of the subject, as we have said, clearly the accident belongs; while if it does not follow, the accident does not belong. You should establish this by induction. (*The Complete Works*, 114b 35–115a 6)

Using statistics from available botanical compendia, Darwin calculates the number of botanical varieties belonging to species with the greatest estimated population sizes and ranges in hopes of discovering some general pattern in, or connection between, these species: "I thought that some interesting results might be obtained in regard to the nature and relations of the species which vary most, by tabulating all the varieties in several well-worked floras" (*Origin* 45).

To make his calculations, Darwin first divides the species in the compendia into large and small species according to the author's size designations. He then divides the number of species in the large and small groups by the number of varieties that are connected with them to produce an average of the number of varieties for each species, large and small. The results of these tabulations and comparisons reveal that there is a correlation between the size and range of a species' population and the number of varieties recorded for that species (Table 2).

Table 2. Arithmetical Comparisons of the Ratios between Species and Varieties in Large and Small Genera. Reprinted from Charles Darwin, *Charles Darwin's Natural Selection*, Ed. R.C. Stauffer, p. 150. © 1975. Used by permission of the publisher, Cambridge University Press.

VARIATION UNDER NATURE

Table I cont.

	Larger Genera	*Smaller Genera* (including those with single species)	*Genera with a single species*
Centre France: Boreau—Larger Genera with 5 species and upwards, smaller with 4 species and downwards.	113/732 = 154/1000 1.38	84/741 = 107/1000 1.47	19/267 = 721/1000 1.47
Holland: Miquel—Larger Genera with 4 species and upwards, smaller with 3 species and downwards.	22/622 = 35/1000	25/557 = 44/1000	
Germany & Switzerland: Koch—Larger Genera with 7 species, and upwards, smaller with 6 species and downwards	390/2093 = 186/1000 1.72	162/1365 = 118/1000 1.79	32/345 = 92/1000 1.50
Dalmatia: Visiani—Larger Genera with 5 species and upwards, smaller with 4 species and downwards.	164/1007 = 162/1000 1.37	130/899 = 144/1000 1.31	46/290 = 158/1000 1.26
Rumelia: Grisebach—Larger Genera with 6 species and upwards, smaller with 5 species and downwards.	98/1136 = 86/1000 1.45	54/1083 = 49/1000 1.14	12/326 = 36/1000 1.16
Russia, Ledebour (All 4 vols together) Larger Genera with 10 species and upwards, smaller with 9 species and downwards	692/3955 = 174/1000 1.48	307/2407 = 127/1000 1.39	45/475 = 94/1000 1.26
Ledebour—Vol: I separately.	207/1237 = 167/1000 1.42	62/576 = 107/1000 1.32	
——— Vol: II ———	192/1243 = 154/1000 1.56	94/767 = 122/1000 1.35	
——— Vol: III ———	171/905 = 188/1000 1.49	94/595 = 157/1000 1.50	
——— Vol: IV ———	122/570 = 214/1000 1.45	57/470 = 121/1000 1.36	
N. United States. A. Gray—Larger Genera with 5 species and upwards, smaller with 4 sp. and downwards. The two kinds of vars. marked in this work are here classed together.	112/1136 = 98/1000 1.40	65/917 = 70/1000 1.36	32/361 = 88/1000 1.37

Based on this arithmetical comparison on the size of Genera and number of species, Darwin argues:

> In any limited country, the species which are most common, that is abound most in individuals, and the species which are most widely diffused within their own country . . . often give rise to varieties sufficiently well-marked to have been recorded in botanical works. Hence it is the most flourishing, or, as they may be called, the dominant species . . . which oftenest produce well-marked varieties. (45–46)

What Darwin discovers, or confirms, as the result of his calculations, is that the more populous species tend to have a greater number of identified varieties associated with them. This correlation is accounted for by his dynamic theory of natural diversity because a correlation between the size of a population and the development of sub-populations would be expected as larger populations would have more offspring, and, therefore, a greater number of variations to be selected.

The correlation between the calculated size of a species and the number of varieties associated with it supports Darwin's argument that diversity in nature is the result of the production of variations. In order to strengthen the conviction of the audience that this relationship is legitimate and to make the case that the relationship exists at all levels of the taxonomic hierarchy, Darwin predicts that the same relationship will be found between genera and species. To validate this prediction he conducts further calculations and comparisons to assess whether or not the principle holds true at the taxonomic level of genera. He walks his readers through his process of calculating and explains his results:

> If the plants inhabiting a country and described in any Flora be divided into two equal masses, all those in the larger genera being placed on one side, and all those in the smaller genera on the other side, a somewhat larger number of the very common and much diffused or dominant species will be found on the side of the larger genera. (46)

Here, Darwin affirms his quantitative prediction that the same correlation between population size and variation which exists between species and varieties also exists between genera and species.

Strengthened by the predictive power of his model and the accumulating evidence, Darwin makes a point to emphasize the success of his theory in accounting for the patterns revealed by his calculations:

> From looking at species as only strongly-marked and well-defined varieties, I was led to anticipate that the species of the larger genera in each country would oftener present varieties, than the species of the smaller genera; for wherever many closely related species . . . have been formed, many varieties or incipient species ought, as a general rule, to be now forming. . . . On the other hand, if we look at each species as a special act of creation, there is no apparent reason why more varieties should occur in a group having many species, than in one having few. (46–47)

In addition to playing up the success of his prediction, Darwin also challenges special creationists to account for the same results. If the different taxonomic categories did in fact represent unique populations of organisms that shared no relationship with other populations, then what would account for the correlations his comparisons reveal? Though opponents of his theory might argue that these correlations are coincidental, Darwin suggests here that the fit between the patterns he describes in the quantitative data and the process of variation that he proposes in his theory is too good to be coincidental (47).

An analysis of the arguments in the second chapter of *The Origin of the Species* reveals that Darwin employed quantitative comparison and basic arithmetical operations to support his theories of dynamic variation and relation by descent between the different levels of the taxonomic hierarchy. In the opening portion of the chapter, he uses quantitative comparison with precise numerical values to challenge the veracity of the existing paradigm of special creation by revealing that, among experts, there is no clear consensus on the categorization of organisms in nature. Once he has cast doubt on the theory of his opponents and offered his own, he accumulates evidence to support it. With evidence from calculated averages and precise numerical comparison and argument from the commonplace of *the more and the less*, he reasons that a correlation exists between the size of populations and the development of recognized variations within related subpopulations. These connections suggest that there is a relationship of descent between different levels of the taxonomic hierarchy wherein varieties associated with a given species are actually variations of a common ancestor, and so forth, up the taxonomic hierarchy.

Without the support of overt, quantitative comparisons and behind-the-scenes mathematical operations, Darwin's argument for the existence of relation by descent would have been purely speculative. But by using mathematical comparison and a quantitative commonplace, Darwin hoped to establish an ethos of precision and rigor commensurate with the conventions for robust scientific argumentation prescribed by Herschel and Whewell and the values of his target audience of geologists, botanists, and zoologists who were caught up, like himself, in the biogeographical revolution.[8]

Chapter IV: Natural Selection and Calculating Diversity

With the evidence and arguments in place that diversity in nature is the result of the spread of variations through organic populations and that a struggle for existence takes place in nature, Darwin proceeds to describe the details of species formation by natural selection.[9] In the fourth chapter of *The Origin of Species,* arithmetical computation and comparison of ratios help Darwin discover and support new lines of argument about selection and species formation, namely: (1) that the more diversified a group of organisms are the better they will do in their struggle for existence, and (2) that the success of highly divergent organisms in part explains how great degrees of difference come to exist between related species.

In the initial stages of his calculations of the ratios of varieties to species, Darwin divided the total number of organisms he was investigating into "large" and "small" groups and calculated the average number of varieties for each species and species for each genus in these size categories. He then compared the average number varieties calculated for the large and small categories of species and genera to determine whether there was a correlation between the size of a species or genre, and the amount of variation produced. (This is the argument strategy explained in the detailed discussion of chapter two of *The Origin of the Species.*)

These calculations revealed that those genera and species designated "large" had more species and varieties. This evidence supported his conclusion that there was a relationship between the size and range of a population and the number of variations it had. A communication from Sir John Lubbock, the son of Darwin's neighbor at Down in the summer of 1857, however, apprised Darwin that building his case on assumed average estimates of size created problems in establishing rationally defensible comparisons of the relative degree to which larger genera might be better producers of species and varieties. Lubbock suggested that instead of *averages*, Darwin should calculate the *ratios* of varieties to species in large genera and then use them to predict the expected ratio of varieties to species in small genera. Although Darwin was initially skeptical about Lubbock's suggestion, he reworked his estimates with good results:

> I have divided the New Zealand Flora as you suggested, there are 339 species in genera of 4 [species] and upwards, and 323 in genera of 3 [species] and less. The 339 species have 51 species presenting one or more varieties. The 323 species have only 37: proportionately (339:323 :: 51 : 48.5) they ought to have had 48 1/2 species presenting vars.–So that the case goes as I want it, but not strong enough, without it be general, for me to have much confidence in. / I am quite convinced yours is the right way; I had thought of it, but should never have done it had it not been for my most fortunate conversation with you. (Darwin to Lubbock, July 14, 1857)

What Darwin discovered in working out the projected ratios of varieties to species in large and small genera was that, in the case of small genera, there are fewer varieties than expected, or, in the case of larger genera, more varieties than expected. The results of these calculations shifted Darwin's attention away from the correlation between the size and range of a group of organisms at a particular level of the taxonomic hierarchy and the number of subordinate categories of organisms associated with that group, and towards the importance of variety to the evolutionary success of organic populations.

In a letter to Hooker in August of 1857, Darwin's belief in the importance of these calculations to the development of his theory is evident.

> I intend dividing the varieties into two classes, as Asa Gray and Henslow give the materials, and, further, A. Gray and H.C. Watson have marked for me the forms, which they consider real species, but yet are very close to others; and it will be

> curious to compare results. If it will all hold good it is very important for me; for it explains, as I think, all classification, *i.e.* the quasi-branching and sub-branching of forms, as if from one root, big genera increasing and splitting up, etc., as you will perceive. But then comes in, also, what I call a principle of divergence, which I think I can explain. (Darwin to Hooker, August 22, 1857)

This shift of attention towards the importance of the breadth of variation to evolutionary success inspired Darwin to think more carefully about why variation might matter so much in the process of selection. These investigations also led him to develop his *principle of divergence of character*, which responds to a question he and his critics considered a major obstacle to any theory of variation: "How do the small differences that are observable between populations of closely related species and varieties grow into the large differences that we see between genera, families, etc.?" (Browne 74).

Because the calculated ratios showed that greater variety was a hallmark of species and genera with larger populations and ranges, Darwin felt encouraged to explain evolutionary success in terms of variation. He reasoned that when a species reaches equilibrium between its numbers and resources, the only way it could continue to grow is through diversification. Evidence from *The Origin of the Species* suggests this line of thinking:

> Take the case of a carnivorous quadruped, of which the number that can be supported in any country has long ago arrived at its full average. If its natural powers of increase be allowed to act, it can succeed in increasing . . . only by its varying descendants seizing on places at the present occupied by other animals: some of them, for instance, being able to feed on new kinds of prey . . . some inhabiting new stations. . . . The more diversified in habits and structure the descendents of our carnivorous animal became, the more places they would be enabled to occupy. (93)

Using the hypothetical case of the "carnivorous quadruped," Darwin argues that a species' survival and replication can be improved if some of its members can expand, through variation, into ecological niches where there is less competition for resources with other conspecifics. In a separate passage, he extends this reasoning, concluding that because larger genera and species produce greater degrees of variation, they will be better off in the struggle for existence.

> In a large genus it is probable that more than one species would vary. . . . In each [large] genus, the species, which are already extremely different in character, will generally tend to produce the greatest number of modified descendents; for these will have the best chance of filling new and widely different places in the polity of nature. (99)

In addition to making the case that bigger is better, Darwin argues by implication that the breadth of diversity of larger genera and species promotes further diversity and population expansion. This means that larger genera and species will most likely continue to grow and outstrip their smaller counterparts, unless checked by some unforeseen circumstances. The supposed correlation between size and variation assumed in this passage suggests that, for Darwin, bigger genera and species are bigger in the first place *because* they are more diverse. Thus, diversity itself becomes a crucial factor in evolutionary success.

Working from the assumption that diversity is a primary factor in selection, Darwin proposes an answer to the question, "How do small differences between populations of closely related species and varieties grow into large differences?" with his *principle of divergence of character.* He presents this principle somewhat inconspicuously in a breeding analogy at the beginning of the section, "Divergence of Character":

> A fancier is struck by a pigeon having a slightly shorter beak; another fancier is struck by a pigeon having a rather long beak; and on the acknowledged principle that 'fanciers do not and will not admire a medium standard, but like extremes,' they both go on . . . choosing and breeding birds with longer and longer beaks, or with shorter and shorter beaks. (92)

This analogy acts as jumping-off point for launching the reader into the hypothetical, "natural" example of the carnivorous quadruped discussed earlier. The "fancier" who selects which features to propagate becomes a conceptual ground for "nature." The large and small beak sizes, and the "acknowledged principle" that "fanciers do not admire a medium standard," prepare the reader to understand and accept the importance of variation as a principle of selection.

Despite these similarities, one important aspect of the analogy—the fancier's continuous choice of the shorter and longer beaks—is not carried over. This small exception deserves attention because, with it, Darwin makes a case for his *principle of divergence of character.* The principle is as follows: If we assume that variation is a primary factor in evolutionary success, and if we assume that for each generation the most varied individuals will be selected (i.e., the "fancier," aka "nature," will consistently avoid the "medium standard"), then it must be the case that, over time, qualities (like beak size) which begin relatively close to each other will be steadily pushed to extremes until members which were once part of the same population are so transformed that they will be perceived as completely different types of organisms. Thus, the answer to the question, "How do small differences between populations of closely related species and varieties grow into the large differences?", is that small differences get magnified into larger ones because selection favors variety over similarity.

An examination of the fourth chapter of *The Origin of Species,* as well as peripheral textual materials, reveals the importance of ratios in focusing Darwin's attention on the centrality of variation to evolutionary success and in supplying him with the inspiration for his *principle of divergence of character.* By working out the ratios for the number of species that might be expected in a large-sized genera, and the number of varieties that might be expected for a given species, Darwin discovered that the actual ratio of variations in smaller groups of organisms is smaller than the predicted ratios. This discrepancy encourages Darwin to focus his attention on variation and to conclude that the degree of variation is an important selection factor. Further, it leads him to argue that the continual selection of highly variable organisms explains how related organic forms can become so radically different that they form different species, and over time. to break off to form separate genera (*Origin* 100–101*).*

The Success and Failures of Darwin's Mathematical Arguments with His Audience

While it is possible to get a glimpse of Darwin's behind-the-scenes efforts to make his conclusions rigorous by employing quantitative data and using mathematical procedures to compare his data, it remains to be determined whether using mathematical arguments had any positive effect on the audience's opinion about the validity of these conclusions. This evaluation is particularly difficult given that *The Origin of the Species* was only an abstract of Darwin's arguments, and therefore lacked the full articulation of the evidence—particularly the mathematical evidence—presented in the manuscript for *Natural Selection*. To get a sense of the effect of the mathematical argument in both of its manifestations, this chapter concludes with an examination of comments and reviews by readers who were aware of the full scale of mathematical evidence, and those who were not.

Because the manuscript of *Natural Selection*, which contained Darwin's data and calculations, had to be trimmed for publication, *The Origin of the Species* included only a few references to his mathematical reasoning. Consequently, readers and reviewers of the text, unless they were extremely diligent in their probing of Darwin's arguments, were not aware of the mass of quantified data and numerous arithmetical calculations that supported his conclusions. Many not-in-the-know reviewers who commented on Darwin's method seemed to support John Herschel's view that Darwin's theory of evolution and natural selection were the laws of "higgledy-piggledy" because there was no clear, inductive path from experiment or observation to the conclusions Darwin had established (Darwin to Lyell, December 10, 1859).[10]

This point is taken up at length, for example, by British mathematician and geologist William Hopkins in *Fraser's Magazine*. In his review of the Darwin's argument, entitled "Physical Theories of the Phenomena of Life," Hopkins argues that the main fault in Darwin's reasoning is that he asserts general laws about selection and variation without the necessary empirical proof that would lead up to them inductively:

> What we require is some proof of fundamental propositions, founded on observation or experiment, as above stated. Till this or something equivalent to it be done, we maintain that no theory like Mr. Darwin's can have a real foundation to rest upon, or can ever secure the general convictions of philosophers in its favor. (80)

A similar criticism of Darwin's method is leveled by the renowned geologist Adam Sedgwick in the *Spectator*. In his review of *The Origin of the Species*, Sedgwick challenges the validity of Darwin's claims on the grounds that they have not been drawn inductively:

> I must in the first place observe that Darwin's theory is not *inductive,*–not based on a series of acknowledged facts pointing to a *general conclusion,*–not a proposition evolved out of the facts, logically, and, of course including them. To use an old figure, I look on the theory as a vast pyramid resting on its apex and that apex a mathematical point. (qtd. in Hull 160)

In the final line, Sedgwick's description of Darwin's theory as a vast pyramid resting on its apex represents an unstable epistemological system balanced precariously on a single point. Instead of moving inductively from a large, secure base of reliable, empirical evidence towards a theory describing that evidence, Darwin, in Sedgwick's estimation, had begun with theory and left the evidence to be discovered by others. Curiously, instead of referring to the top of the pyramid as deductive, natural law, he refers to it as having "a mathematical point." This would make sense on the grounds that in Victorian science—where Newton was hailed as the paragon of scientific achievement—a mathematical description of natural phenomena was considered the apex of scientific effort. However, it seems odd in a critique of Darwin's work to accuse him of resting on a mathematical point when, in fact, the mathematical reasoning that he used in the text was obscured. In fact, the mathematical quality of Darwin's argument is neither mentioned again in Sedgwick's critique in the *Spectator,* nor does it appear when Sedgwick makes the same critique elsewhere.[11] Although it remains something of a mystery as to why Sedgwick would use the phrase, "a mathematical point," it seems likely that this was a general reference to the fact that mathematical laws had a privileged position at the top of the Victorian pyramid of knowledge, rather than a critique of Darwin's work as relying too much on mathematical reasoning.

Though it is impossible to say whether the publication of the complete manuscript of Darwin's *Natural Selection* would have made his conclusions more palatable to Hopkins and Sedgwick, members of his audience—such as Watson, Gray, and Hooker, who had access to the quantitative evidence and were sympathetic to the goals of biogeography—were more amenable to his conclusions. In a letter from Watson on November 8, 1855, for example, Watson describes his feelings about the legitimacy of Darwin's method of statistical argument as well as his reservations about the conclusions Darwin draws from them:

> You may correctly believe that I *do* take an interest in your investigations about the numerical evidences or indications of variability. There seems to be this difference between us, that you regard the proportions as approximating more towards evidences of a general law, while I see them only as fainter indications of it. (H.C. Watson to Darwin, November 8, 1855)

In these lines, Watson offers support for Darwin's method of using quantitative evidence and arithmetical calculations to make his case, but complains that Darwin has been too quick to make grand claims from his data. Unlike Hopkins and Sedgwick, however, he is not ready to dismiss Darwin's conclusions as without merit or evidence. Instead, he believes they are credible but at a lower level of certainty than he believes Darwin is intending.

Like Watson, Gray seems to embrace Darwin's mathematical approach; however, unlike Watson, he seems to be more convinced that Darwin's conclusions are supported by his evidence and calculations. In his lengthy review of *The Origin of the Species* in the *American Journal of Science and Arts* (1860), Gray clearly states his acceptance of Darwin's mathematically derived argument—that the correlation between the size and range of a genus' population and the number of species associated with it—is evidence of a relation of descent:

> That the most flourishing and dominant species of the larger genera on an average vary most (a proposition which can be substantiated only by extensive comparisons, the details of which are not given); and . . . that in large species the genera are apt to be closely but unequally allied together. . . . The fact of such association is undeniable; and the use which Darwin makes of it seems fair and natural. (170)

Of particular interest in this section of Gray's review is the comment in the parentheses, in which Gray alludes to the extensive quantitative comparisons that he knows Darwin has undertaken to make his case. By alluding to them in his review he assures his readers that Darwin has been diligent in gathering and comparing data to support his conclusions, and that the reviewer has, based on his knowledge of Darwin's arithmetical calculations, found these proofs compelling.

Like Gray, Hooker also supports Darwin's conclusions and his mathematical approach to developing his arguments in *The Origin of the Species*. Hooker voices his belief in his introduction to volume three of *The Botany of the Antarctic Voyage the Flora of Tasmania* (1860), which is previewed in the *American Journal of Science and Arts.* In this introduction, Hooker admits to his own stunning conversion by Darwin and Wallace's theory from his previous belief that species were fixed:

> In the introductory essay to the *New Zealand Flora,* I advanced certain general propositions as to the origin of species . . . amongst others was the still prevalent doctrine that these are . . . created as such, and are immutable. In the present essay I shall advance the opposite hypothesis, that species are derivative and mutable; and this chiefly because . . . every candid mind must admit that the facts on which he has grounded his convictions require revision since the recent publication . . . of the ingenious and original reasonings and theories of Mr. Darwin and Mr. Wallace. (2)

Some of the "ingenious and original reasoning" that persuaded Hooker came from Darwin's arithmetical calculations with quantified data. This fact is evidenced by a statement later on in the introductory essay where Hooker described his agreement with Darwin's *principle of divergence of character,* the calculations of which Darwin communicated to Hooker in August, 1857: "Mr. Darwin addresses another principle in action amongst living organisms as playing an important part in the origin of species, viz., that the same spot will support most life when peopled with very diverse forms" (Darwin to Hooker, August 22, 1857).

Though Hooker supported Darwin's *principle of divergence of character,* he was skeptical that his calculations were reliable because they depended on distinctions between varieties and species that had no natural basis. He voiced his concern in a letter to Darwin in March, 1858:

> Botanists do not attach that *definite* importance to varieties that you suppose, they do not treat large & small genera equally & similarly, & the sum of inequalities thus produced tends to make the species of small genera look more invariable than of big. . . . So my dear Darwin do not be in a hurry with your conclusions. (Hooker to Darwin, March 14, 1858)

In these lines, Hooker warns Darwin, as Herschel and Whewell caution their readers, that quantitative argument is robust only to the degree to which the mathematical evidence can be linked to natural reality.

Despite their misgivings, Hooker, Watson, and Gray, all well-respected botanists, accepted Darwin's arguments, and in some cases, completely reversed their views on the origin of species. The persuasion of these experts is a testament to the strength of Darwin's strategy to use quantified evidence and arithmetical calculations to make his case for *dynamic variation* and the *principle of divergence of character.*

Conclusion

With the quantitative data and arithmetical calculations hidden, it is no surprise that historians and rhetoricians of science, like some of Darwin's contemporaries, found no evidence of mathematical argumentation in Darwin's work. However, to say that Darwin could not have used mathematics to make his arguments ignores the evidence in the letters, notebooks, manuscripts, and the text of *The Origin of the Species.*

A close examination of his reading habits and notebooks reveals that Darwin was dedicated very early in his scientific endeavors to developing a mathematical approach to the study of biological phenomena. His communications with Watson, Gray, and Hooker, as well as in the construction of his arguments in the manuscript of *Natural Selection*, give evidence that he put these beliefs into practice.

In the construction of his arguments, Darwin was also trying to follow the best practices of making scientific knowledge. Though he offered no formulae for describing selection or the *principle of divergence of character*, he did depend on quantified data to make his case, and he did attempt to use arithmetical calculations to discover or supply evidence for the existence of patterns in natural phenomena. These practices are essential steps up the ladder of quantitative induction outlined by Herschel and Whewell.

Ironically, despite his best efforts to construct rigorous hypotheses based on quantified data and arithmetical comparison, Darwin's ideas only "seem" mathematical to Pearson, and appear to have no quantitative basis to others because of an unexpected turn of events. With the emergence of Wallace's paper, Darwin was forced to abandon his plans to publish a detailed discussion of his arguments and evidence. Instead, he put out an abstract of his work. In the process of summarization, much of the quantitative evidence and mathematical calculations supporting his theory were left out. As a consequence, Darwin's work was vulnerable to the very charges of methodological weakness that he had assiduously attempted to avoid by using quantitative evidence and arithmetical methods of comparison. Through a close, rhetorical examination of Darwin's arguments, as well as a careful assessment of his reading habits, ideas, and correspondences, it is possible to recover Darwin's efforts to construct quantitative theories of *dynamic variation* and the *principle of divergence of character* according to the best practices of Victorian scientific argument. These efforts suggest that more credit needs to be given to Darwin as a pioneer of mathematical argument in the study of organic phenomena who's every idea "fit itself to mathematical definition" and demanded statistical analysis.

4 Hidden Value: Mendel, Mathematics, and the Case for Uniform Particulate Inheritance

> *In the study of evolution progress had well-nigh stopped. . . . Such was our state when two years ago it was suddenly discovered that an unknown man, Gregor Johann Mendel, had, alone, and unheeded, broken off from the rest—in the moment that Darwin was at work—and cut a way through.*
>
> —William Bateson

> *It is clearly important to test these remarkable statements by a careful study of the numerical results. . . . It seems to me that by neglecting these precautions some writers have been lead to overlook the wonderfully consistent way in which Mendel's results agree with his theory.*
>
> —Frank Weldon

Despite Darwin's efforts to provide a robust theory of variation and evolution, important pieces of the puzzle were still missing. He had, for example, no reasonable theory for the mechanism of inheritance that could address critics such as Fleeming Jenkin, who argued that if variations were small and needed to be accumulated regularly over vast amounts of time, they would become blended away over repeated generations of cross-breeding within the normal population. Although Darwin did develop his theory of pangenesis in which he argued that different characteristics of the parents were transferred to their off-spring by gemmules (small, trait carrying particles in the blood), this theory was unproven and could not explain how characters could remain sufficiently uncompromised over time to accumulate (Hull 308-09).

Unknown to him and the majority of natural philosophers at the time, the problem of character stability was being worked out by a monk laboring carefully over selected and organized samples of peas. The monk, of course, was Gregor Mendel, and the carefully planned series of breeding experiments that he undertook from 1856 to 1864 provided the evidence for what we now recognize as the first, modern theory of genetic inheritance.[1] Although Mendel had described in "Versuche über Plfanzen-Hybriden" ("Experiments in Plant Hybridization," 1865) the basic pattern of inheritance of seven separate traits in peas, the value of his work on heredity was not recognized until his findings were simultaneously "rediscovered" by three researchers in 1900: Hugo de Vries, Carl Correns, and Erich von Tschermak. Examining Mendel's arguments for his hereditary model, and his failure to persuade biological researchers to accept it, provides fertile ground for investigating the rhetorical dimensions of mathematical argument in science. A close analysis of his argument suggests that he employs an analogy between heredity and the mathematical principles of probability and combinatorics to establish his natural laws of inheritance. Further, Mendel's theory and experimental design reveal that mathematics also serves as a source of invention for arguments rather than just a tool for describing natural phenomena. In addition, Mendel's failure to persuade his audience with mathematical arguments reveals that arguing with mathematics in science is a complex and uncertain affair wherein mathematics doesn't always enjoy the status of *demonstrative reasoning.* Instead, it sometimes has the standing of argument in the domain of "the credible, the plausible, and the probable"—in other words, the rhetorical (Perelman and Olbrechts-Tyteca 1).

Method

To illuminate these rhetorical dimensions of Mendel's mathematical argument, this chapter will rely on a number of different methodologies, including historical, textual, and audience response analysis. With the help of historical background information about the epistemological and ontological commitments of eighteenth and nineteenth century European, continental hybridists, a conceptual framework for understanding the prevailing opinions about heredity and the role of mathematics in its exploration during that period will be constructed. These opinions help establish the tenuous nature of Mendel's arguments and the manner in which considerations other than analytical ones might affect the status of mathematical argument as a legitimate method for making claims about nature.

In addition to constructing the theoretical context in which Mendel made his argument, this chapter also engages in a close reading of his arguments in "Experiments in Plant Hybridization." Careful scrutiny of the text provides evidence that mathematics was a source for invention in developing Mendel's conclusions about heredity and in the design of the experiments to reach those conclusions. It also reveals the centrality of analogy to the prosecution of his arguments. Finally, an examination of reader responses to the text provides evidence for why Mendel's work may have been rejected or generally ignored. Analysis of the only critique Mendel received of his work and his responses to that critique suggest that he had more faith in the validity of his mathematical argument than his audience did.

Hybridism before Mendel

To understand why Mendel's use of the mathematical principles to make arguments about heredity generated conflict with his audience, it is necessary to know something about his audience's intellectual commitments. A brief exploration of the theories of and methods for investigating heredity of Joseph Gottlieb Kölreuter and Carl Friedrich von Gärtner—arguably the most influential theorists in nineteenth century continental hybridism—offer some perspective on the agreements about hybridization and hybridism shared by Mendel's contemporaries.

Although experimentation in hybridization was considered as far back as 1694 by Rudolph Jacob Camerer, in his work, *Über das Geschlecht der Pflanzen* (*About the Gender of Plants*), [2] systematic research in hybridization and the crossing of plants considered to be different varieties and species began in the mid-eighteenth century with the work of Joseph Gottlieb Kölreuter (1733–1806) and was carried on in the nineteenth century by Carl Friedrich von Gärtner (1772–1850). In general, these researchers dealt with questions regarding the variation or fixity of natural forms and the physiological process by which either variety or homogeneity was transferred from one generation to the next. Specifically, they wanted to know: (1) how much parents contributed of their characters to their offspring, (2) whether their contributions were similar, (3) how the characters of the parents were carried into the offspring, and (4) how they developed.

Joseph Gottlieb Kölreuter

Joseph Gottlieb Kölreuter was born on April 27, 1733, in the Sawbian village of Sulz in the Black Forest region of southwest Germany. His early hybridization experiments were conducted in a variety of places, including his hometown, in the garden of Achatius Gärtner in the nearby town of Calw, as well as in St. Petersburg, Berlin, and Leipzig (Roberts 35–36). The four-volume work Kölreuter developed from these investigations included *Vorläufige Nachricht von einigen das Geschlecht der Pflanzen betreffenden Versuchen und Beobachtungen* (Preliminary Report about Some Observations and Experiments Regarding the Gender of Plants) and its three *Fortsetzungen* (*Continuations*), which were written between 1761 and 1766. These volumes were considered the central, authoritative texts on hybridization by nineteenth century plant hybridists. In these works, Kölreuter sets down many principles that inform Mendel's research, though Mendel's conclusions differ in important ways.

In his earliest experiments, Kölreuter aimed to test Camerer's conclusion that both male and female plants contributed material to the reproductive process, and to understand in more detail the manner in which and the degree to which they contributed. Towards this end, he created hybrids by crossing a total of thirteen genera and fifty-four species over a six-year period, including his famous crossing of two different varieties of tobacco (*Nicotina paniculata* and *Nicotina rustica*), the first recorded experiment in artificial plant hybridization (Roberts 36). Hybridization provided an ideal method for investigating parental contributions to offspring because the contrast between the parents' characteristics provided an easy way to determine the extent and nature of their contributions. If hybrid characters favored one or the other of the parents, Kölreuter could assume that the generative force of that parent was stronger than the other.

What he found in the crosses of tobacco was that the hybrid always seemed to be a form intermediate between the two parents. The outcome of these crosses led Kölreuter to reconfirm Camerer's position that plants do have different sexual components that contribute material to the reproductive process. It also suggested to him that not only do both the male and female parents contribute to the production of offspring, but also that each of their contributions is approximately equal and blended, as the hybrid seemed to be a perfect intermediate of maternal and paternal characteristics (Roberts 42–43).

In addition to providing answers about the role of parents in reproduction, Kölreuter's experiments also led him to an uneasy contemplation of the stability of species. As a proponent of essentialism inherent in the Christian doctrine of special creation, which held that each species was a unique and unalterable act of creation, Kölreuter was troubled by his ability to create a hybrid type, an alteration of specific form. This uneasiness led him to perform further back-crossing experiments to establish the degree to which species might be changed. The results were mixed. Although some hybrids were completely infertile, a fact which confirmed that divine barriers were in place to protect the integrity of species, others produced viable offspring. To understand the stability of these offspring as new species, Kölreuter conducted further experiments in which he back-crossed hybrids with their original parent species. These crosses resulted in plants that were, for the most part, similar to one of the two original species, a few that were related to the form of the other parent species, and a few that displayed faint, mixed traces of both parents. The reversion of most plants to their parental forms in these experiments led Kölreuter to conclude that this, too, was evidence of a barrier in nature against the mixing of specific essences (Mayr 644).

From the results of these back-crossing experiments, Kölreuter developed a theory of inheritance, which maintained that each species or variety involved in hybridization expressed itself in different degrees, depending on the strength of a particular species' essence. (This concept of dominance appears in Mendel's work, though it is a characteristic of a particular genetic trait rather than a species' essence.) In the initial stages of reproduction, Kölreuter argued, the essences of the two species "commingled" to create an "intermediate material" for the hybrid plant (Roberts 47). During the plant's process of development, however, the commingled essences are separated and expressed in what would now be called the phenotype, the outward physical appearance of the plant, according to the original strength of their essence.[3] The result is a hybrid offspring (F1) which resembles both the parental species in keeping with the degree of vitality of their essences:

> All the movements and changes, which from the embryo to the time of flowering, take place. . . . All aim at gradually liberating that compound material upon which they are based, and at dividing it again into the two original ground materials; or, to speak more properly, to bring these latter themselves into a complete, and, especially from the one side, into masses of unlike size than were demonstrated from the preceding reproduction. (qtd. in Roberts 47–48)

Kölreuter's hybrid experiments provided the conceptual framework for most nineteenth century researchers' beliefs about, and investigations into, the hybridization process. Most importantly, they took up his ideas that (1) hybrids represented an equal or near equal commingling of parental species' essences, (2) hybrids of species are generally infertile with crosses of closely related forms, showing a higher degree of fertility, (3) in the offspring of fertile hybrids, a few of the progeny reverted to the parental hybrid form while most reverted to the grandparent forms in the original cross, and (4) the degree to which traits are expressed in the physical characteristic of hybrids is directly related to the strength or amount of the species' essences that are commingled.

In addition to adopting his theories about the heredity process and the character of the hereditary substance, later researchers also co-opted many of Kölreuter's methods for conducting experiments, drawing conclusions, and graphically presenting results. In the bulk of his work, Kölreuter crossed what were assumed at the time to be distinct species (e.g., *Nicotiana paniculata* X *Nicotiana rustica*), though the subsequent fertility of his crosses led him to believe in some cases that certain species might be better described as varieties. From these carefully recorded crosses, he gathered both qualitative and quantitative data on the appearance of hybrids and their relations to their parents from which he drew conclusions about the manner and degree to which traits were passed on from parents to offspring by their expression within the offspring.

The results of Kölreuter's crossing experiments were presented graphically using natural language, brackets, and the symbols for male (♂) and female (♀). In Figure 1 from the third continuation (*Dritte Fortsetzungen*) in 1776, for example, he uses these graphical conventions to describe experiments for crossing of *Nicotiana paniculata* and *Nicotiana rustica.* Under the heading "Experiment 23" (XXIII Vers.), Kölreuter uses visuals to communicate that in the first round of crosses

Figure 1. Visuals for Kölreuter's crossings of *Nicotiana paniculata and Nicotiana rustica in Vorläufige Nachricht von einigen.* (Leipzig, 1893; 196).

[48] **XXII. Vers.**

Nicot. rust. ♀.
rust. ♀.
} ♀.
panic. ♂.
Nicot. ♀.
panic. ♂.
♂.
panic. ♂.

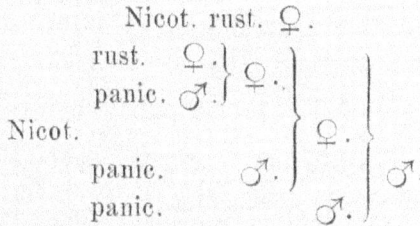

Von diesem Versuche wurden sechs Pflanzen erzogen. Ich konnte zwischen ihnen und den einfachen aus der rust. ♀ und panic. ♂ erzeugten Bastarten keinen merklichen Unterschied finden.

XXIII. Vers.

rust. ♀.
Nicot. ♀.
panic. ♂.
rust. ♀.
} ♀.
panic. ♂.
Nicot. ♀.
panic. ♂.
♂.
panic. ♂.

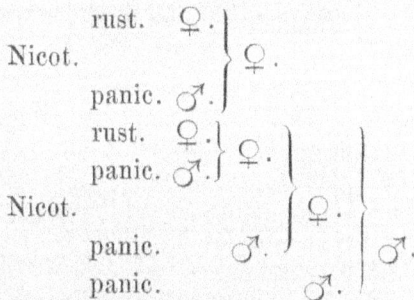

Ich erzog hievon drey Pflanzen. Eine derselben war ihrer ganzen äusserlichen Anlage nach dem in der zweyt. Forts. § 16. S. 73. etc. beschriebenen Bastart im ersten aufsteigenden Grade sehr ähnlich, und hinterliess viele, aber ganz leere Kapseln. Die zwo übrigen hatten etwas weniger Aehnlichkeit mit der panic. als die erstern, und setzten nur sehr wenige, ziemlich spitzige und ebenfalls ganz leere Kapseln an. Man sieht hieraus, [49] dass sie mit mehrern andern dergleichen Bastarten im ersten aufsteigenden Grade übereingekommen sind.

in the experiment he mated two female *N. rustica* plants with two male plants of the species *N. paniculata.* The results of the crosses were two female offspring. Of the two female offspring, Kölreuter bred the second one twice with two different non-hybrid male *N. paniculata*. This double breeding is indicated visually by the use of long and short brackets and an empty space at the top of the second bracket. The short and long brackets indicate that two different males were mated with the female hybrid, and the empty space at the top of the second, long bracket signals the reader that the hybrid female from the first breeding was also crossed in the second breeding with a different male. In the first breeding, a male hybrid offspring was produced, in the second, a female.

While Kölreuter relies on visual conventions to provide readers with the details of his experimental set-up and procedure, he uses natural language descriptors to describe the characteristics of the hybrids that result from the crosses. In the text below the visual description of "Experiment 23" in Figure 3, he explains that the result of the first crossing, presumably the one at the top of the chart, resulted in a hybrid that was an intermediate between the two parents and produced no seeds.[4] The second and third crossings, however, resulted in hybrid offspring that were intermediate but slightly less similar to the male *N. paniculata* fathers.[5]

In addition to using visuals and natural language to describe his crossing experiments, Kölreuter also employed tables of quantitative data to compare various physical features of the parents and their hybrid offspring. This application of quantification represented an important component of Kölreuter's arguments about hybridization. By juxtaposing the measurements of parent plants and their offspring in a chart format he was able to show that, in some cases, hybrid crosses were intermediate forms of their parents, and in other cases, one parent's character dominated.

In Table 3, "Measure of Comparison" (Vergleichungsmaass), for example, Kölreuter presents his readers with comparative measures for different physical features of parents and hybrid offspring in his crossing experiments with *N. rustica* and *N. paniculata*. Listed in the far left column are the features measured, such as "length of the entire flower from the base of the stem to the edge of the flat, spread-out, five-lobed blossom" (Länge der ganzen Blume von dem Grunde der Blumenröhre an bis zu dem flach ausgebreiteten und in fünf Einschnitte abgetheilten Blumne rande), "width (or rather length) of the outer rim itself of the blossom" (Breite (oder vielmehr Länge) des abstehenden Blumenrandes slebst), and "entire length of the flower tube" (Ganze Länge der Blumenröhre) (65).

Table 3. Kölreuter's quantitative comparisons of hybrid offspring and non-hybrid parents for a variety of features. Source: *Vorläufige Nachricht von einigen* (65).

Vergleichungsmaass.

	Nicot. rust.	rust. ♀ ☿ panic. ♂	panic.	rust. ♀ ☿ α. panic. ♂ } ♀ peren. ♂	rust. ♀ ☿ β. panic. ♂ } ♀ peren. ♂	peren.
Länge der ganzen Blume von dem Grunde der Blumenröhre an bis zu dem flach ausgebreiteten und in fünf Einschnitte abgetheilten Blumenrande	7‴	$9\frac{2}{3}$‴	1″, $1\frac{1}{2}$‴	1″, 3‴	1″, 8‴	2″
Länge des Blumenkelchs: von seinem Grunde an bis an die Spitze des längsten Einschnitts	$5\frac{1}{2}$‴	$5\frac{1}{2}$‴	$3\frac{2}{3}$‴	$7\frac{1}{2}$‴	$7\frac{1}{2}$‴	10‴
Die Blume ragt über die Spitze des längsten Kelcheinschnitts heraus . .	$1\frac{1}{2}$‴	4‴	$9\frac{1}{2}$‴	$6\frac{1}{2}$‴	1″, $\frac{1}{2}$‴	1″, $2\frac{1}{2}$‴
Grösste Breite von einem Ende des ganzen Blumenrandes bis zum andern, quer über die Blume gemessen . .	$7\frac{1}{3}$‴	$5\frac{1}{3}$‴	$4\frac{1}{3}$‴	9‴	$10\frac{1}{3}$‴	1″, $2\frac{1}{2}$‴
Breite (oder vielmehr Länge) des abstehenden Blumenrandes selbst .	$2\frac{1}{3}$‴	2‴	$1\frac{1}{4}$‴	$3\frac{1}{3}$‴	5‴	6‴
Durchmesser der Blumenröhrenöffnung zwischen dem Rande . . .	$2\frac{3}{4}$‴	$2\frac{1}{4}$‴	$1\frac{2}{3}$‴	3‴	$3\frac{2}{3}$‴	$3\frac{2}{3}$‴
Durchmesser des Blumenröhrenbauchs unter dem Rande	$3\frac{1}{2}$‴	3‴	$2\frac{1}{4}$‴	4‴	$3\frac{1}{4}$‴	4‴
Ganze Länge der Blumenröhre . .	$6\frac{2}{3}$‴	9‴	1″, $\frac{2}{3}$‴	1″, $1\frac{1}{4}$‴	1″, 7‴	1″, 10‴
Länge des engen Grundes der Blumenröhre	$1\frac{1}{2}$‴	$2\frac{1}{4}$‴	3‴	$4\frac{1}{4}$‴	6‴	$6\frac{1}{2}$‴
Länge der Staubfäden	4‴	$5\frac{1}{2}$‴	$8\frac{1}{3}$‴	$9\frac{1}{2}$‴	11‴	1″, $3\frac{2}{3}$‴
Länge des Stils	$4\frac{1}{2}$‴	7‴	$11\frac{1}{4}$‴	$11\frac{1}{2}$‴	1″, $4\frac{1}{2}$‴	1″, $6\frac{1}{3}$‴
Länge des Eyerstocks, die gelblichte Substanz mit eingeschlossen.	$1\frac{1}{3}$‴	$1\frac{1}{2}$‴	$1\frac{1}{2}$‴	2‴	3‴	$3\frac{1}{2}$‴
Durchmesser des Eyerstocks über der gelblichten Substanz.	$1\frac{1}{3}$‴	1‴	$\frac{3}{4}$‴	$1\frac{1}{2}$‴	$1\frac{1}{3}$‴	$1\frac{2}{3}$‴

In the first three columns of measurements, Kölreuter records from left to right the values for the defined characters for the female parent *N. rustica,* the hybrid offspring, and the male parent *N. paniculata* used in the crossing experiment. An examination of the values for the first physical feature on the chart—"length of the entire flower from the base of the stem to the edge of the flat, spread-out, five-lobed blossom"—reveals that the value of this feature for the hybrid is larger than the sum of parents, which indicates in this case that heredity is additive. For the next feature, however, "length of the calyx, from its base to the tip of the longest [deepest] incision" (Länge des Blumnekelchs: von seinem Grunde an bis an die Spitze des längsten Einschnitts), the measure for the hybrid is exactly that of the female *N. rustica* parent. This similarity suggests that this feature in the hybrid was exclusively influenced by the mother. Finally, the measurements of the third feature, "the [amount the] flower projects above the tip of the longest [deepest] calyx incision" (Die Blume ragt über die Spitze des längsten Kelcheinschnitts heraus), presents evidence that the characteristics of the two parents blend almost equally, as the value for the hybrid offspring is a little less than half the combined value of the parents.

By arranging character measurements on the chart so they can easily compared by his readers, Kölreuter illuminates the variety of ways in which heritable features can combine, and provides evidence for his characterization of heredity as being the consequence of a commingling of specific essences of various potencies. However, aside from taking this initial step in the process of quantitative induction, Kölreuter did not make any effort to develop a mathematical analogy describing the abstract patterns with which traits combined.

Carl Friedrich von Gärtner

In the German village of Calw, in the same garden where Kölreuter had conducted some of his hybridization experiments, one of the greatest supporters of his work, Carl Friedrich von Gärtner, undertook his own research program of crossing hundreds of species and varieties in the early nineteenth century (Olby 49). A physician by trade and son of a distinguished botanist Joseph Gärtner (1732–1791), Carl Gärtner did twenty-five years worth of meticulously executed and recorded hybrid crossing experiments.

The initial motivation for carrying out these experiments is not clear. In fact, Gärtner's results might have passed into obscurity if not for a prize offered by the Dutch Academy of Sciences at Haarlem for essays answering the query, "What does experience teach regarding the production of new species and varieties, through the artificial fertilization of flowers of the one with pollen of the other, and what economic and ornamental plants can be produced and multiplied in this way?" (qtd. in Roberts 167). In October of 1835, Gärtner learned of the prize and sent the academy a brief sample of his work. Because he had not formally compiled all of his results, he asked for an extension to complete his research. The extension was granted, and two years later he presented them with a two-hundred-page memoir that, in 1849, was published as a monograph under the title, *Versuche und Beobachtungen über die Bastarderzeugnung im Pflanzenreich* (*Experimentation and Observation of the Creation of Hybrids in the Plant Kingdom*). In this monograph, Gärtner details nearly 10,000 separate crossing experiments among 700 species belonging to 80 different genera of plants (Roberts 168).

From these experiments Gärtner concludes, as Kölreuter had, that the essences of the different species are commingled in the hybrid. However, unlike Kölreuter, who believed in only one true hybrid form, Gärtner divides hybrids, based on the nature of their commingling, into three categories: (1) *intermediate*, (2) *commingled*, and (3) *definite*. Hybrids that expressed in equal amounts pure traits from each of their parent species were identified as members of the *intermediate* class of hybrids because "a complete balance occurred of both fertilizing materials, in respect to either mass or activities" (168). In this definition, Gärtner describes what he understands are two important factors that affect the degree of expression of a particular species' essence in the hybridized offspring: (1) the mass or amount of a particular species' essence present in an offspring, and (2) the activity or potency of the essences present.

What Gärtner argues is that the presence of more or less mass or activity on the part of one species' essence results in the expression of a lesser or greater number of that species' characters in the phenotype of the hybrid offspring. Like Kölreuter, Gärtner believed that different aspects of the two species' essences were expressed in different parts of the phenotype. However, he makes no attempt to experiment on specific traits because he believes that traits are expressions of different facets of the unified fabric of a species' essence rather than a mosaic of separate units. He describes the antithesis of essences and their expression in different domains of the phenotype when he writes:

> In the formation of simple hybrids, as in sexual reproduction in general, two factors are active. This unlikeness of activity, flowing from the specific differences of species, expresses itself through the more pronounced or the weaker manifestation of the individual parental characters in the different parts of the hybrid. (qtd. in Roberts 171)

Whereas the *intermediate* form is created by an exact balance of contributions from one parent or the other, the *definite hybrid*, according to Gärtner, is one, "among which the resemblance of a hybrid to one of its parents . . . is so marked and preponderating that the agreement with the one or with the other is unquestioned" (169). Here Gärtner argues that when the mass or activity of one of the species is greater than the other, the pure features of the predominant species are preponderant in the form of the offspring. Though Gärtner's concept of preponderance appears similar to Mendel's notion of dominance, it is different in the sense that Gärtner has no recessive counterpart. There is no sense that the reproductive material from the non-dominant parent is somehow preserved to be passed on to later offspring.

Whereas *intermediate* and *definite hybrids* exhibit clear affinities with either or both of their parents, Gärtner's third category of hybrids, *commingled hybrids*, include plants in which the characters expressed are not exactly like either of the parents. This lack of clear affinity is attributed by Gärtner either to outside causes which affect the reproductive materials before they commingle, or to the influence of the two essences of the parents on each other. In his explanation of *commingled hybrids*, Gärtner writes about the former:

> Now this and now that part of the hybrid approaches more to the maternal or to the paternal form, whereby, however, the characters of the parents, in their transference to the new organism, never go over pure, but in which the parental characters always suffer a certain modification. (qtd. in Roberts 168)

In his discussion of back-crosses, Gärtner suggests that the modification in the pure characters might have something to do with the presence of the essences of the different parent species. However, the modification is not one in which different contributions from each of the parents are mixed to create a new character that is not purely one or the other. Instead, the presence of each of the species' essences has a modifying influence on the manner in which the dominating features of the other is expressed:

> [The fundamental ground material of the hybrid] behaves differently in the second and in the further stages of breeding, where, on account of the different nature of the two factors of the hybrids in the succeeding fertilizations, an altered, shifting, variable direction in type formation enters into the arising varieties (169).

In their work, Gärtner and Kölreuter supplied tentative answers to important questions about the processes of reproduction in plants. They explored trait combination and theorized that the essences of two species commingled and separated out, with each species expressing itself in a particular domain of the offspring. They inquired into the phenomenon of dominance, and concluded that the amount or activity of specific essences affected the extent to which the characteristic of a parental type was expressed in its hybrid offspring. Finally, Gärtner questioned the difference between hybrid offspring and their parents, and hypothesized that either outside influences or influences from the presence of the essence of the other species affected the pure expression of characters.

Mendel's understanding of the fundamental precepts of plant hybridization theory and his primary argument for the stability of traits came from the works of Kölreuter and Gärtner. However, the impetus for his research was the result of questions that still lingered in the minds of hybridists regarding the quality and quantity of the contribution that each different parental species made to the hybrid. Though Gärtner and Kölreuter had theorized about why hybrid forms had a mixture of characters from either parent which varied in different degrees, researchers still sought a clearer understanding of what accounted for such differences in the expression of characters.

Against the Grain: Mendel and his Contemporary Audience

Variation and Fixity of Species

One of the great mysteries surrounding Mendel's arguments in "Experiments in Plant Hybridization" has been why, given the paper's centrality to the development of our modern discipline of genetics, was it so thoroughly ignored by the biological researchers who were the first to read it? The answer suggested in this section is because Mendel's theory was contrary in almost every way to the ontological and epistemological commitments of the botanists, hybridists, and other biological researchers who would have read it. The evidence which follows—both from Mendel's work and the work of those researchers who received copies of his work—reveals that Mendel accepted aspects of Gärtner and Kölreuter's work that most researchers rejected, and rejected elements they accepted.

One of the most important differences between Mendel's position on heredity and the ontological commitments of his contemporaries was that Mendel, like Gärtner and Kölreuter, believed in the fixity of species and the non-transmutability of hereditary traits, while most other biological researchers circa 1865 were participating in something of a cultural revolution against the two great hybridists on these positions. During the 1840s and 1850s, for example, a radical biological movement developed in the work of Franz Unger, Karl Nägeli, and others that rejected Gärtner and Kölreuter's position that species were fixed categories, and embraced Jean Baptist Lamarck's (1744–1829) position in *Philosophie Zoologique* (1809) that species could change over time.

Franz Unger (1800–1870) was Mendel's professor of physiology and plant paleontology in Vienna from the fall of 1852 to the spring of 1853. In his weekly "Botanical Letters" column in the *Vienna Times* (*Wiener Zeitung*) in the winter of 1856, he writes, "Who can deny that new combinations arise out of this permutation of vegetation, always reducible to certain law combinations, which emancipates themselves from the proceeding characteristics of the species and appear as a new species" (qtd. in Henig 63).

These words of Mendel's instructor are companions to the sentiments of Carl von Nägeli, the only person known to have both read and responded to Mendel's work. In Nägeli's *Individuality in Nature with Especial Reference to the Vegetable Kingdom* (*Die Individualitat in der Natur mit besonderer Berucksichtigung des Pflanzenreiches)* (1865), he writes:

> Like natural phenomena in general, species cannot persist in complete repose. Just as the offspring of the first individual were a little different from that individual, so also must the

> germs which engendered them diverge to some extent from those out of which they themselves originated. A process of change must be perennially at work, and this change cannot fail, in the end, to bring about the disappearance of the species or its transition into another. (qtd. in Iltis 186)

Because Mendel's professors and other important members of the biological community were swept up in this new movement challenging the fixity of the species, some historians believe that Mendel's experiments had been designed for the purpose of unraveling the mysteries of species transformation as well. In *Monk in the Garden,* for example, historian Robin Henig argues,

> Maybe Mendel set out, in his pea experiments, to confirm his idea of "perennial" progressive change as a driving force for the appearance of new species from the old. If he designed his experiments correctly, he could lend empirical support to the theories of two botanists he had come most to admire: Nägeli and Unger. (64)

Henig and other historians' position that Mendel's purpose for undertaking his investigation as an exploration of the transformation of species has some support in the text of "Experiments."[6] In the opening, Mendel himself suggests that, at least in part, his experiments had been undertaken to contribute to the discussion regarding the variability of the species.

> It requires indeed some courage to undertake a labor of such far-reaching extent; this appears, however, to be the only right way by which we can finally reach the solution of a question the importance of which cannot be overestimated in connection with the history of the evolution of organic forms. (2) [7]

Though there seems to be some compelling evidence supporting the view that Mendel undertook his experiments to explore the variability of species, other evidence from the text of "Experiments" suggests that, in fact, Mendel was pursuing the opposite conclusion: that species are in most cases *not* variable. Towards the end of "Experiments," for example, Mendel appeals to Gärtner's conclusion that species, even cultivated ones that have been bred to vary, cannot become so variable that they lose their fixity and can be transformed into another species: "But nothing justifies the assumption that the tendency to the formation of varieties [in cultivated plants] is so extraordinarily increased that the species speedily lose all stability, and their offspring diverge into an endless series of extremely variable forms"[8] (32).

In addition to espousing positions in the text that suggest skepticism towards the possibility of organic transformation through variation, Mendel is also very particular about framing his work in the context of hybridists who support the inviolability species. In the "Introductory Remarks" section, Mendel recognizes Gärtner, Kölreuter, and Wichura as precursors of his work, all of whom believed in the fixity of the species.

> The striking regularity with which the same hybrid forms always reappeared whenever fertilization took place between the same species induced further experiments to be undertaken, the object of which was to follow up the developments of the hybrids in their progeny.
>
> To this object numerous careful observers, such as Kölreuter, Gärtner, Herbert, Locoq, Wichura, and others, have devoted a part of their lives with inexhaustible perseverance. (1) [9]

In this section of the introduction, Mendel describes his work using terms such as "regularity" (Regelmässigkeit) and "reappearance" (wiederkehrten) of traits during hybridization, not by their variability or disappearance. These word choices suggest the focus of his investigation is on the inviolability of traits rather than their mutability. Further, the absence of mentioning the work and ideas of Unger, Nägeli, and Darwin suggest that Mendel does not envision his work in the tradition of studies about the transformation of species.[10]

Essence

Whereas Mendel finds himself in opposition to the ontological commitments of important continental natural researchers by standing with Gärtner and Kölreuter on the topic of the fixity of species, he opposes them by taking a position against the classical hybridists on the subject of specific essences. This difference is highlighted in Mendel's discussion of his materials and methods, particularly in his criteria for choosing experimental subjects.

In the first two sections following the introduction, Mendel provides his readers with a careful accounting of and rationale for both the materials and the methods of his experiments. In the first section, "Selection of Experimental Plants" (Auswahl der Versuchspflanzen), he offers readers a list of the character traits that he intends to observe in his experiments. There are seven characters, including: (1) the difference in the form of the ripe seeds (smooth/wrinkled), (2) the difference in the color of the seed albumin (pale yellow, bright yellow, or orange/intense green), (3) the difference in the color of the seed coat (white/grey, grey/brown, leather brown, or violet), (4) the difference in the form of the ripe pods (smooth/wrinkled), (5) the difference in the color of the unripe pods (light to dark green/vividly yellow), (6) the difference in the position of flowers (axial along the main stem/terminal at the end of the stem), and (7) the difference in the length of the stem (long, 6–7ft. and short, ¾-1½ ft.) (5–6).

Kölreuter, Gärtner, and Nägeli believed that all of the characters, or describable features of a plant, were part of a species essence, and therefore had to be considered *in toto* when studying heredity. However, this list reveals that Mendel assumed that features such as height, seed color, etc. could be studied independently and that information about the process of heredity could be gained from their individual study. This assumption required a whole new way of thinking about organisms as well as novel methods for investigating heredity.

A brief overview of the introduction and methods and materials section of Mendel's text, in the context of prevailing theories about and methods for experimenting with heredity, suggests that Mendel's research ran counter to the prevailing ontological commitments of important natural researchers whom he attempted to persuade with his work. On the one hand, Mendel's efforts opposed the research programs of natural philosophers like Unger and Nägeli because of its solidarity with the classical hybridists on the fixity of the species. On the other hand, Mendel stood against Gärtner, Kölreuter, and his contemporaries on the issue of essence. From a rhetorical perspective, the contrariness of Mendel's position on so many important issues raises questions: "Why did Mendel adopt such a radically different position from his contemporaries?" And "On what grounds did he believe he could persuade them to accept such a radical theory of heredity?"

Mathematics as a Source of Invention

Mathematics in Mendel's Education

To understand both the contrariness of Mendel's position as well as his confidence that he could persuade others to accept it, it is necessary to examine his background, the sources of invention for his arguments, and the arguments themselves. A brief foray into his personal history and the development of his argument suggests that Mendel firmly believed in the power of mathematics to understand nature, and drew on it to design his experiments. It also suggests that these beliefs were important factors in his confidence that his efforts would be embraced by his colleagues.

Mendel's belief that mathematics had the power to unveil the shrouded mysteries of nature likely came from his own experiences in university as both a student of mathematics and of physics. While attending the University of Vienna from 1851–1853, Mendel enrolled in courses with the great physicist Christian Doppler, and with the renowned mathematician Andreas von Ettingshausen. In fact, Mendel's interest in mathematics and the physical sciences was so pronounced that it made up, according to Alain Corcos and Floyd Monhagan, at least half of his academic schedule (*Gregor Mendel's Experiments* 24).

Along with learning the subject matter of mathematics and physics, Mendel also had an education in how to apply the knowledge in these fields to investigations of the natural world. He was chosen by Doppler as an assistant demonstrator at the university's physical institute. In this capacity, he learned how to perform experimental demonstrations of various physical phenomena. This practical education in the methodology of experiments was to have a profound impact on how he conducted his own work with hybridization.

In addition to the methodology of experimentation, Mendel was also introduced to the use and importance of mathematics as a tool for predicting and modeling physical phenomena. At the time of his studies, statistics and probability had become an important part of this area of knowledge (Orel 31). Orel suggests that Mendel might have learned something of the subject from Doppler, who had published a mathematical textbook in 1844 with a chapter on, "combinatorial theory and basic principles of probability calculation" (31).[11]

Orel also suggests that Mendel could have gained some understanding of the theoretical application of probability to natural philosophy from the work of Joseph Johann von Littrow (1781–1840), with whose small volume, *Probability Calculation as Used in Scientific Life* (1833), Mendel was also familiar. In the text, Littrow discusses the application of Laplace's theory of probability to all phenomena, arguing that, "the relationships of all phenomena in nature seem at first completely random, but the greater the number of those phenomena that are considered, the closer they approach to certain constant relationships" (qtd. in Orel 32). Orel explains that Mendel relied on the book as a resource for compiling meteorological records, and for making meteorological forecasts of the weather in Moravia (32).

Mathematics as a Guide for Experimentation

Mendel's mathematical education and his experience as an assistant demonstrator of physical experiments introduced him to the commitments of the physical sciences, including the importance of mathematical principles and formulae in the construction of robust knowledge about nature. In "Experiments in Plant Hybridization," he carries these commitments with him in the creation and design of his biological experiments, using the law of probability as an analogical source to guide his understanding of, and arguments for, the hereditary process.

As was explained in chapter two, analogy is a comparison between dissimilar constituents for the purposes of delighting the audience, clarifying obscure ideas, or aiding the intellect in discovering truths about nature. Specifically, analogies are considered to have two constituent parts: the *phoros* and the *theme.* In the *New Rhetoric,* Perelman and Olbrechts-Tyteca explain that the *phoros* is the part of the analogy with which the audience is familiar. This constituent of the analogy provides a structure, value, and/or meaning by which the unknown or unvalued *theme* can be understood or characterized (372–73).

In Mendel's analogy, the identification of the *phoros* and *theme* seems straightforward. The hereditary process would be the *phoros* because it represents a concrete, biological process that could be observed and measured, whereas the classical model of probability would be the *theme* because of its abstractness. In his article, Mendel makes every effort to give his audience the impression that this is indeed the structure of his analogy. He spends the first half of the paper describing the results of his empirical experiments and uses the second half to build them into mathematical laws. Careful examination of his experimental procedure, however, reveals that, in fact, the mathematics came first and served as the *phoros* for the analogy, while the model of the hereditary process came second, defined by the parameters of probability.

In order to understand the mathematical model's influence in Mendel's experimental design, it is helpful to consider first the details of the analogy between the *phoros* (probability) and the *theme* (hereditary process), and then analyze the text for evidence of their influence on his experimental design. Whereas the (bat/eye) (soul/understanding) analogy discussed in chapter two is a simple analogy involving just two sets of corresponding constituents, the (probability)/(hereditary process) analogy is a rich or complex analogy involving multiple connections between the *phoros* and the *theme.*

In the *phoros* of probability, there are three features that guide the development of Mendel's experimental design:

1a. In every case where there are two trials with two possible outcomes (such as heads or tails), there are only four possible results (HH, TH, HT, TT).

2a. Establishing the relative frequency of outcomes can only be reliably attained from a large number of trials.

3a. The values of outcomes should be arrived at by the assessment of aggregate, average values.

Analogs of these three features are present in the following aspects of Mendel's experimental design.

1b. Traits used in experimentation must be stable/pure and remain that way over many crossings.

b. Traits used in experimentation must be clear binaries.

2b.Many crosses must take place in order to describe the pattern of inheritance.

3b. The average of the outcomes of all crosses for a particular trait in a particular generation must be compared.

Unlike most hybridists—whose experiments were relatively simple and involved (1) selective breeding of different species, (2) prophylactic measures for limiting accidental fertilization, and (3) strict record keeping of results—Mendel's experimental method was much more detailed, requiring additional criteria for selecting the traits to be bred and therefore setting the standards for the amount of data that was to be gathered and how it was to be assessed. These novel specifications can be linked to the parameters of mathematical probability, suggesting that Mendel was going out of his way to design his experiments to test the hypothesis that the heredity process followed the law of probability.

Evidence that mathematical probability guided Mendel's investigations can be found in his selection criteria of traits for experimental crossing. In the first part of the materials and methods portion of the text, "Selection of Experimental Plants," Mendel describes the three basic criteria he used in choosing and controlling experimental subjects:

1. [That they] Possess constant differing characters.
2. That the hybrids of such plants must, during the flowering period, be protected from the influence of all foreign pollen, or be easily capable of such protection.
3. The hybrids and their offspring should suffer no marked disturbance in their fertility in the successive generations. (2)

Of these criteria, the first is unique to Mendel's work because it seeks to ensure that crossing begins with features that are pure or constant. This unique requirement is essential because it makes it possible for Mendel to test the hypothesis that the reproductive process in plants functions according to the laws of probability. In the simplest case involving two simultaneous, probabilistic trials of subjects with binary outcomes (HT), there can only be four possible results (HH, TH, HT, TT). The predictability of these results is predicated in part on the condition that the possible outcomes of each flip will remain the same (H or T). If, for some reason, they were to transform to something else (say Q), then no reliable probability could be hypothesized using the standard model. In order to guarantee that his hypothesis could be judged against the model for probabilistic outcomes, Mendel needed to ensure that his traits were pure and remained stable from one generation to the next.

In addition to maintaining the purity of traits, Mendel also had to ensure sufficient contrast between the characters so that the outcomes of crossings could be accurately judged as either one trait or another. Towards this end, Mendel purposely elected to hybridize only plants with highly distinctive traits. These included traits such as the difference in the form of the ripe seeds (smooth/wrinkled), the difference in the color of the seed albumin (e.g., pale yellow, bright yellow, or orange/intense green), etc. He underscores the importance of the distinctness of these physical features when he writes:

> Some of the characters noted [such as the size and color of flowers] do not permit a sharp and certain separation since the difference is of a "more or less" nature, which is often difficult to define. Such characteristics could not be utilized for the separate experiments; these could only be applied to characters which stand out clearly and definitely in the plants. (4)

By eliminating features that might appear graded, Mendel again avoids an issue of comparability between his results and the outcomes deduced from probabilistic models. He recognizes that if the results of his trials could not be certified as having one particular outcome or another, he could never make the case that trait combination and distribution followed the patterns prescribed by the law of probability.

In addition to selecting plants whose traits made them eligible test subjects for proving his hypothesis, Mendel also chose to follow particular methods of data collection and analysis that were ideally suited to uncovering probabilistic patterns. These methods included making a large number of trials and assessing the results of trials in aggregate rather than on a case-by-case basis. The importance of doing a large number of trials to assess the probability of an event was a well-established tenet of mathematical probability by the beginning of the nineteenth century. It was laid out by Jakob Bernoulli in his limit theorem nearly two centuries before. The theorem states that the calculated *a posteriori* probability of an event, p, gets closer to the true *a priori* probability of an event, P, the greater the number of trials, n, that are conducted (Chatterjee 168).

In his description of the outcomes of his experiments in the section, "The First Generation from the Hybrids," Mendel seems to reference this law when he discusses the importance of his sample size in eliminating aberrations that might hide the true underlying stochastic patterns:

> As extremes in the distribution of the two seed characters in one plant . . . were observed. . . . These two experiments [in which ten plants were observed] are important for the determination of average ratios, because with a smaller number of experimental plants they show that very considerable fluctuations may occur. (10)

In addition to doing enough trials to eliminate abnormal cases, it is also necessary to establish the underlying pattern of probability through aggregate, average values rather than looking for it by comparing individual cases, such as the patterns in the outcomes presented in a given pea pod or plant. In a second passage mirroring the one previously quoted, Mendel explains the importance of this consideration when assessing probabilistic outcomes:

> It remains purely a matter of chance which of the two sorts of pollen may fertilize each separate egg cell. For this reason the separate values must necessarily be subject to fluctuations and there are even extreme cases possible. . . . The true ratios of the numbers can only be ascertained from an average deduced from the sum of as many single values as possible. (25–26)

Examination of the textual evidence suggests that Mendel relies on mathematical probability as a *phoros*, a well-understood set of principles, with which to gain a foothold in understanding the hereditary process. This relationship is evident in the designs of his experiments, which pave the way for testing this analogy by ensuring that (1) a sufficiently large number of trials are made, (2) the outcomes of all crosses for a particular trait in a particular generation must be compared and averaged, (3) the traits being crossed must be pure, and (4) these outcomes can be clearly distinguished from one another as binaries. Mendel's imposition of these strictures on his experimentation suggests that he was guided by rather than led to a belief that heredity operated according to the principles of probability. They also explain why he embarked on a theory of heredity whose assumptions ran counter in many ways to those of his contemporaries.

From Analogy to Reality

Whereas Mendel uses the first five sections of "Experiments" following the introduction to establish the analogy between probability and heredity, he employs the remainder of the text trying to erase it. This process of unifying the *phoros* and *theme* into a single, reciprocally descriptive system represents the ultimate aim of science: the creation of a mathematically describable, natural law. In *The New Rhetoric*, the authors identify erasure as a final step in the scientific process:

> Analogy finds a place in science, where it serves rather as a means of invention than as a means of proof. If the analogy is a fruitful one, theme and phoros are transformed into examples or illustrations of a more general law, and by their relation to this law there is a unification of the fields of the theme and the phoros. This unification of fields leads to the inclusion of the relation uniting the terms of the phoros and of the relation uniting the terms of the theme in a single category, and, with respect to this category, the two relations become interchangeable. There is no longer an asymmetry between theme and phoros. (Perelman and Olbrechts-Tyteca 396)

The conflation of the biological with the mathematical begins in the section titled, "The Subsequent Generations from the Hybrids" (Die weiteren Generationen der Hybriden) in Mendel's *Experiments in Plant Hybridization*. In the opening lines of this section, Mendel states that his aim is to establish a law of heredity based on the results of his experiments and observations described in the first half of the text. He writes, "The proportions in which descendents of the hybrids develop and split up in the first and second generations presumably hold for all subsequent progeny" (13).[12]

The first step Mendel takes in making the case that the relationships he has witnessed in his experiments represent a law of nature is to transform the dominant and recessive character traits and their quantitative relationship to one another into a single formula: $A+2Aa+a$.

> If A be taken as denoting one of the two constant characters, for instance the dominant, a, the recessive, and Aa the hybrid form in which both are conjoined, the expression
>
> $$A+2Aa+a$$
>
> shows the terms in the series for the progeny of the hybrids of two differentiating characters. (14) [13]

After the initial presentation of this mathematical expression describing the ratio of traits in a hybrid generation, Mendel, realizing perhaps that his audience might object to his mathematical transformation, attempts to win their consent by showing that his expression provides deductive proof for reversion, a quality of hybrids identified by Kölreuter and accepted as commonplace by hybridists:

> The observation made by Gärtner, Kölreuter, and others, that hybrids are inclined to revert to the parental forms, is also confirmed by the experiments described. . . . If an average equality of fertility in all plants in all generations be assumed, and if, furthermore, each hybrid forms a seed of which one-half yields hybrids again, while the other half is constant to both characters in equal proportions, the ratio of numbers for the offspring in each generation is seen in the following summary. (14) [14]

Following these comments, Mendel provides the reader with a chart in which his expression for the distribution of traits is applied hypothetically over multiple generations:

Generation				Ratios		
	A	Aa	a	*A*	: *Aa*	: *A*
1	1	2	1	1	: 2	: 1
2	6	4	6	3	: 2	: 3
3	28	8	28	7	: 2	: 7
.				.		
n				$2^n - 1$	2	$2^n - 1$ (14)

The result of this application reveals that over each successive generation the pure parental forms do increase dramatically while the hybrid forms remain constant. This suggests, mathematically, that the expression that Mendel is arguing in favor of predicts the widely supported observation of reversion.

In addition, the chart also facilitates the reader's acceptance of the law-like nature of his findings by logically transitioning them from empirical data to the rational mathematical relationship between them. The chart begins with the fixed ratio that he has uncovered through observation of the experimental outcomes of his pea breeding experiments. It then moves the reader through successive, hypothetical generations to reveal a dynamic but regular pattern of change. This movement eventually leads them to a general mathematical formula which, if they have accepted the regular pattern of trait distribution in the empirical data and its regular action on successive generations, they are compelled to accept. This final formula, $2^n - 1: 2: 2^n - 1$, is described by Mendel as *the law of development* (Entwicklungs-Gesetz).

With a mathematically describable law stipulating the regularity of the appearance of a single pair of traits over many generations in hand, Mendel moves on to prove that this law applies to a whole suite of traits in an individual organism, so long as those traits breed true before crossing. In the next section of the text, "The Offspring of Hybrids in Which Several Differentiating Characters are Associated" (Die Nachkommen der Hybriden, in welchen mehrere differirende Merkmale verbunden sind), he accomplishes this task by using accumulated empirical evidence from controlled hybridization experiments to move his audience further from empirical results toward abstract, mathematical manipulations that offer deductive support for his position.

Mendel begins his argument with natural language descriptors, but quickly moves to reduce them to capital and lower case letters. This reduction allows his readers to understand what traits he is discussing without his having to write out those traits every time he discusses them.

Expt. 1 —	*AB*, seed parents;	*ab*, pollen parents;
	A, form round;	*a*, form wrinkled;
	B, albumen yellow.	*b*, albumen green. (15) [15]

After he has established his system of notation, Mendel gradually moves from natural language descriptors of his data towards mathematical expressions of the combinatorial relationships between the traits he discusses. This progression is apparent in the manner in which the results of the second, third, and fourth-round crosses are described with increasing abstraction. The first round of dihybrid self-crosses (*F1*) are described using numbers and natural language phrases in a fashion similar to the descriptions of the data in the first three sections of the results portion of the text, except that they are listed rather than presented within the syntax of a natural language sentence.

315 round and yellow,

101 wrinkled and yellow,

108 round and green,

32 wrinkled and green. (15) [16]

In the next section describing the second round of dihybrid self-crosses (*F2*), Mendel introduces abstraction by including the letter symbols for the traits alongside the natural language descriptions.

38 had round yellow seeds	*AB*
65 round yellow and green seeds	*ABb*
60 round yellow and wrinkled yellow seeds	*AaB*
138 round yellow and green, wrinkled yellow and green seeds	*AaBb* (16)[17]

Following this initial chart, however, he offers a more orderly accounting of the results of the second round of dihybrid crosses without much natural language.

38 plants with the sign *AB* [[18]]

35 " " " " *Ab*

28 " " " " *aB*

.

.

. (16)

As the number of possible results increases, the amount of natural language in the text decreases. This transition is warranted because it permits Mendel to control the clarity of his results as they increase in complexity. At the same time, however, the movement away from natural language affords him the opportunity to lead his readers from the concrete results of his experiments towards the abstract, mathematical expressions he wishes to establish as general descriptors for the pattern of inheritance. In fact, by the end of his description of the results of the dihybrid crosses, Mendel has reduced all of the relationships into a mathematical analog summarizing the result of the crosses.

> Consequently the offspring of the hybrids [F2]—if two kinds of differentiating characters are combined therein—are represented by the expression
>
> *AB+ab+Ab+aB+2AaB+2Aab+2ABb+2aBb+4AaBb.* (17) [19]

It is not hard to imagine that this neat expression of the final result of the dihybrid crosses would be cumbersome for readers if it were written in natural language. Mendel's expression, on the other hand, does a great deal to make the results accessible while at the same time leading readers towards two important conclusions. First, by transitioning through a series of forms moving from raw data expressed in natural language towards increasingly abstract mathematical representation, Mendel encourages his reader to see in the concrete, semi-variable empirical data the expression of a fixed, regular natural law. Second, Mendel proves for a second time the validity of his first law stated in the expression, *A+2Aa+a*, by deriving it again deductively from the expression, *AB+ab+Ab+aB+2AaB+2Aab+2ABb+2aBb+4Aa Bb*.

> This expression [*AB+ab+Ab+aB+2AaB+2Aab+2ABb+2aBb+ 4AaBb*] is indisputably a combination of series in which the two expressions for the characters *A* and *a*, *B* and *b* are combined. We arrive at the full number of classes of the series by the combination of the expressions:
>
> *A+2Aa+a*
>
> *B+2Bb+b.* (17) [20]

In the last portion of this section, Mendel follows the same procedures in order to prove that character traits maintain a constant ratio of distribution in trihybrid crosses. He begins again by presenting the raw data and then transitions toward a more abstract presentation as the data becomes more copious. Finally, he presents the relationship between the data as a mathematical expression from which he derives, once again, the basic formulae for the ratio of traits for a given series of crosses.

After accumulating the evidence from both the dihybrid and trihybrid crosses, and showing that the ratio of characters can be mathematically deduced from their results, Mendel, drawing on his knowledge of combinatorics, concludes:

> There is therefore no doubt that for the whole of the characters involved in the experiments the principle applies that *the offspring of hybrids in which severally essentially different characters are combined exhibit the terms of a series of combinations, in which the developmental series for each pair of differentiating characters are united.* (19) [21]

Now that Mendel has established, through the accumulation of inductive proof and through mathematical deduction, that the result of the combination of different traits in dihybrid and trihybrid crosses is always simply the combination of the separate, expected ratios for each of the individual characters, he concludes that, "*the relation of each pair of different characters in hybrid union is independent of the other differences in the two original parental stocks*" (19).[22] With this statement, Mendel issues his formal challenge to traditional theories of inheritance that emphasized a homogeneous species essence by arguing—based on his ability to mathematically extract each of the ratios independently from the combined expression of traits—that character pairs are separate and non-interacting.

Once he has proved deductively that characters assort independently, Mendel uses the mathematical law of combinations to predict the number of non-variable traits that should appear if all the possible combinations are made. This prediction is the final move by which he attempts to erase all asymmetry in the analogy between the empirical and the mathematical. He begins by providing the formulae for calculating, (1) the number of combinations possible for a given number of character pairs, (2) the number of individuals that will occur for all possible combinations, and (3) the number of trait unions which will remain constant in subsequent generations of self-crossing.[23] Collectively, Mendel refers to these formulae as *The Law of Combination of Different Characters.*[24]

> If n represents the number of the differentiating characters in the two original stocks, 3^n gives the number of terms of the combination series, 4^n the number of individuals which belong to these series, and 2^n the number of unions which remain constant. (19) [25]

Based on the laws of combinatorics, the number of forms which remain constant can be calculated by plugging the number of trait pairs (n) into the expression 2^n. Although he does not supply the data of all of the possible combinations for seven crosses, Mendel argues that his crossings have yielded 128 constant unions, the exact number predicted by his law for seven character pairs. This fact, he argues, gives practical proof, "*that the constant characters which appear in the several varieties of a group of plants may be obtained in all the associations which are possible according to the laws of combination, by means of repeated artificial fertilization*" (19).[26] In other words, no matter what trait or variety is crossed, a predictable number of forms will always remain constant. Thus, Mendel is able to deduce and to provide empirical evidence to support his primary claim for the fixity of species.

After establishing credibility with his readers that he is describing a law of nature by making successful predictions about the fixity of specific traits, Mendel provides them, in the final section of the second half of "Experiments," with a complete description of the hereditary process from fertilization to expression, based on these laws. He begins deductively by appealing to the principles of probability to make the case that the results of self-crossing a hybrid plant should lead to a situation in which every possible combination of the dominant and recessive traits is expressed because each of these characters is equally represented in the pollen and egg cells. This argument from the principles of probability is important because it supports both the veracity of his character ratios and his claims of equal qualitative and quantitative contributions of parents in fertilization.

> It remains . . . purely a matter of chance which of the two sorts of pollen will become united with each separate egg cell. According, however, to the law of probability, it will always happen, on the average of many cases, that each pollen form, *A* and *a,* will unite equally often with each egg cell form *A* and *a.* (25) [27]

What Mendel describes here under the title, "the law of probability" (Regeln der Wahrscheinlichkeit), is actually a combination of principles. The first is the principle of elementary outcomes, which suggests that given a single, fair, and independent trial, the result of any given outcome will be 1/number of possible outcomes. Since there are two possible outcomes in this situation—a dominant trait (*A*) or recessive trait (*a*)—there is always exactly a .5 percent chance that a particular variable will appear. Because, however, Mendel is interested not only in the individual probability of the appearance of one trait or another, but also in the probability of the appearance of either in conjunction with another, it is also necessary to calculate the probability of the two independent, equiprobable events occurring together. This can be easily done by referring to Pascal's *Arithmetical Triangle*, which permits the calculation of the binomial coefficient, or the possible successes of a particular outcome given a certain number of trials (*n).*

In Mendel's model, we must imagine that each joining of a dominant with a recessive trait involves the same probability as the flipping of two separate coins, with the possible pairings either both dominant (*A*), both recessive (*a*), or one (*A*) and one (*a*). Accepting that the probability(*P*) of each event is ideally .5, and that the number of trials (*n*) taking place in each pairing is 2, Pascal's triangle reveals that the probability distribution is: *AA* = .25, *aa* =.25, and Aa or aA = .50 (Gonick and Smith 77). These calculations reveal that the average distribution of these traits in any given generation should be ideally 1:1:2, the exact ratio that Mendel describes in his expression A+2Aa+a.

Starting with the assumption that the action of fertilization is random and independent in each case, as in a probabilistic trial, Mendel presents his reader with a step-by-step walkthrough of the process of trait combination. He presents this information visually, interspersing mathematical symbols into the description and giving it the air of a mathematical proof.

He begins by describing the traits in the soon-to-be-crossed parents as equally manifested in the egg and the pollen of each. His choice of the ordering and juxtaposition of traits reinforces his position on the similarity of their contributions.

> The pollen cells (die Pollensellen) *A+A+a+a*
>
> The egg cells (die Keimzellen) *A+A+a+a* (25)

Once he has defined the starting scene of the process, he then proceeds to a diagram describing the action of crossing using arrows to indicate the direction of the coupling of the characters. This visual is, of course, based on the ideal probable outcome of the situation rather than what might really happen in any given crossing of two characters.

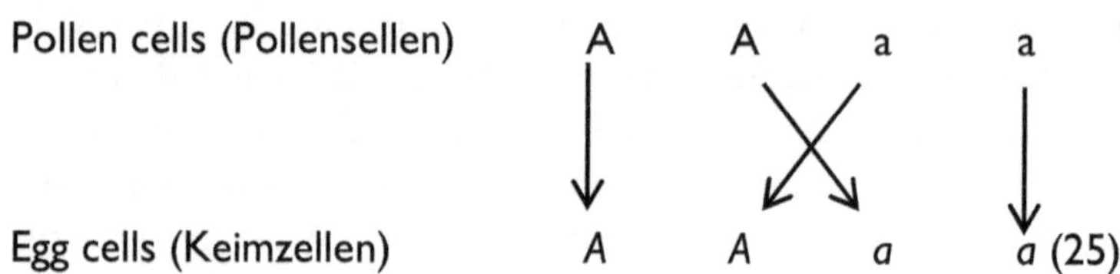

Once Mendel is finished showing the action of cross-pollination combining two traits into a single plant, he symbolically presents his readers with the results of the combinations.

$$\frac{A}{A} + \frac{A}{a} + \frac{a}{A} + \frac{a}{a} \quad (25)$$

In this expression of the results, the symbols for division and addition play important roles. The horizontal lines between the top and bottom row of letters reinforces for the reader the idea that the traits in a new organism are both together, but at the same time, exist as separate, particulate, and non-interacting entities. The addition symbol, "+", between the sets of traits signals that the traits are part of a series of results obtained by the same procedure. In neither of these cases do the symbols actually entail the usual mathematical relations for which they are used; however, they suggest that Mendel is describing the steps of the process in terms of the steps of a mathematical proof.

That these symbols represent separation and membership is fairly evident in Mendel's final transformation in which he restates the relationship of these traits in the set in terms of his familiar 1:2:1 ratio expression.

$$\frac{A}{A} + \frac{A}{a} + \frac{a}{A} + \frac{a}{a} = A+2Aa+a \qquad (26)$$

By starting with what he believed to be established premises from his previous proof, and from established principles in the field of probability, Mendel ends this major line of argumentation in the results section with a step-by-step deduction of the ratio describing the distribution of traits in a given generation. This procedure allows Mendel to visually represent for his readers all of the pieces of the logical puzzle he is constructing, and the mathematical neatness with which they all fit together. In doing so, he suggests not only the deductive veracity of his conclusion, but also the reciprocity between the initially asymmetrical parts of his original analogy between nature and mathematics.

Critical Responses to Mendel's "Experiments in Plant Hybridization"

The previous sections of this chapter have established that (1) Mendel and his audience had opposing commitments, (2) Mendel's commitments are in part the result of his belief in the reciprocity between mathematics and nature, and (3) Mendel dedicates his efforts in "Experiments in Plant Hybridization" to making the case for this reciprocity by first establishing an analogy between heredity and the mathematical principles of probability, and then erasing it. What has yet to be supplied is direct evidence that Mendel's ontological commitments inspired resistance in his audience towards his arguments, and that Mendel's belief in the reciprocity between mathematics and heredity gave him confidence that his conclusions were legitimate despite their controversial nature. By examining the dissemination of Mendel's work, the critique from his audience, and the response Mendel made to that critique, evidence of the audience's resistance to, and Mendel's confidence in, his hereditary model can be established.

The route to publication for Mendel's work was the most common one for naturalists at the time: through the proceedings of the local, natural history/philosophy society to which they belonged. The results of his experiments were first presented in two, one-hour lectures in February and March of 1865 at meetings of the Brünn Natural History Society. The reaction that Mendel got from his colleagues foreshadows, to some degree, the reception that his paper would get from the wider scientific audience. According to Mendel's own account of his lectures, his findings were considered unorthodox and, therefore, controversial.

> I knew that my results were not easy to harmonize with contemporary science, and that in view of this publication of an isolated experiment might be doubly dangerous. . . . I did my best to institute control experiments, and for that reason at the meeting of the Society for the Study of Natural Science [Iltis's translator's version of the society's title, "Naturforschenden Vereines in Brünn"] I described my experiments with Pisum. As was only to be expected, I encountered very various views, but, so far as I know, no one undertook a repetition of the experiments. (qtd. in Iltis 180)

Even though the society's members probably found Mendel's results controversial at best and forgettable at worst, his complete lecture was published in the *Proceedings of the Brünn Natural History Society.*[28] Undaunted by the lukewarm response from his peers in Brünn, Mendel requested forty reprints of his article in order to share his findings with important figures in hybridization and related fields, hoping they might recognize and support what he believed were extremely important results.

Of the forty reprints, twelve are known to have been sent out to some of the most important figures studying variation, evolution, cytology, botany, hybridization, and reproduction at the time Mendel was writing. Notable recipients of Mendel's reprints included scientific luminaries such as Martinus Beijerinck, a Dutch biologist and co-discoverer of viruses; M.J. Schleiden, the establisher of cell theory; Carl von Nägeli, a well-respected cytologist who worked with Schleiden; and Franz Unger, Mendel's botany professor from Vienna. Although six other copies are known to exist, there is no evidence who their original owners are (Henig 142–46).

Among German and Austrian biologists and cytologists there is a common thread of connection that suggests why Mendel believed they would be the most receptive audience for his work. During the 1840s and 1850s in Germany and Austria, there was a small but persistent trend in research away from the qualitative and descriptive methodologies of traditional hybridists and toward the more disciplined experimental/mathematical approaches used in the physical sciences. According to Robin Henig, this method, described in the work of the cytologists Matthais Schleiden and Theodor Schwann, gained traction under the expression "scientific botany," and found supporters in both Nägeli and Unger (Henig 57). Given the central role that mathematics played in Mendel's arguments, it seems reasonable that Mendel might have believed that his best strategy for getting an audience for his controversial conclusions would have been by appealing to researchers who might sympathize with his method, and thereby his conclusions, no matter how unorthodox.

Of this group of influential and mathematically inclined biological researchers, Carl von Nägeli was the only one known to have fully read, considered, and responded to Mendel's work. In his critique of "Experiments," Nägeli maintains a less than enthusiastic position on his conclusions. In a letter to Mendel dated February 25, 1867, he writes:

> Your design to experiment on plants of other kinds is excellent, and I am convinced that with these different forms you will get notably different results (in respect of the inherited characters). It would seem to me especially valuable if you were able to effect hybrid fertilizations in Hieracium. (qtd. in Iltis 192)

In his response, Nägeli begins by encouraging Mendel to pursue future experiments. Though this is a show of confidence in his methodology, the second part of the sentence challenges the general validity of his results. Nägeli's suggestion that Mendel experiment with *Hieracium,* or hawkweed, gives some clues about why he believes Mendel has overstated his claims. Whereas peas provided the ideal subjects for making the case for non-blended particulate inheritance—they self fertilize and have a standard set of two chromosomes—*Hieracium* is a classic example of blended inheritance; it is a polyploid, an organism with multiple sets of chromosomes that could be very easily construed as blending when crossed. By suggesting that Mendel try to obtain his same results using the same experimental techniques with *Hieracium,* Nägeli wanted to make the point that blended inheritance was a very real phenomenon in nature.

Nägeli's orthodox position on the blended nature of inheritance is expressed in an article titled, "Hybridization in the Plant Kingdom" (Die Bastardbildung im Pflanzenreich), published the very same year as Mendel's "Experiments."

> The rule, however, is that the characters of the father and the mother combine and interpenetrate, whereby a new individual character originates which holds more or less the mean. The way and the manner in which the union occurs cannot be determined in advance. (qtd. in Roberts 96)

In order to defend his unorthodox position on the particulate and non-blended nature of heredity, Mendel took up Nägeli's challenge to experiment with *Hieracium.* The ensuing experiments revealed that indeed the plant did not give the same results as his work on peas. Discouraged by this conclusion, Mendel gave up on hybridization experiments altogether.

In addition to attacking Mendel's work on the grounds that he had ignored the phenomenon of blended inheritance, Nägeli also challenged the scope of his claims, particularly Mendel's position that his mathematical predictions warranted the assumption that his hereditary model was relevant for all organisms. In his critique of the universal applicability of Mendel's results, Nägeli begins with what seems initially as a ludicrous accusation that Mendel had not done a sufficient number of experiments: "It seems to me that the experiments with Pisum, far from being finished, are only beginning" (qtd. in Iltis 191). Although this seems like a ridiculous accusation given that Mendel's conclusions were based on the crosses of nearly 10,000 pea plants, Nägeli's complaints are not that Mendel did not do enough experiments with peas, but that he did not do a sufficient number of experiments with other species. Because his sample is deep but not wide, Nägeli argues that Mendel cannot assume a broad scope of application for his conclusion that traits were inalterable:

> The mistake made by all the more recent experimenters is that they have shown so much less perseverance than Kölreuter and Gärtner. . . . You should, however, try to excel them and in my view this will only be possible . . . if experiments on an exhaustive character are made upon one single object in every conceivable direction. (qtd. in Iltis 191)

Nägeli's comment about the breadth of Mendel's sample may have been related to Mendel's move in his argument to present his empirical findings as the source for a predictable, mathematically describable law of nature. According to Mendel, in his second letter to Nägeli, written April 18, 1867, Nägeli had written him that he, "should regard the numerical expressions as being only empirical, because they cannot be proved rational" (qtd. in Stern and Sherwood 63). The point that Nägeli was trying to make is that Mendel could not rely on rational proof from the principles of mathematics to support his argument for the regularity in the distribution of traits in hybridized peas. In his reply, Mendel defends his use of deductive reasoning and reveals his confidence that his mathematical conclusions justify his general assumptions.

> My experiments with single traits all lead to the same result: that from the seeds of hybrids, plants are obtained half of which in turn carry the hybrid trait (Aa), the other half, how-

> ever, receive the parental traits A and a in equal amounts. . . . Therefore 2Aa+A+a or A+2Aa+a is the empirical simple series for two differing traits. Likewise it was shown in an empirical manner that, if two or three differing traits are combined in the hybrid, the series is a combination of two or three simple series. Up to this point I don't believe I can be accused of having left the realm of experimentation. If then I extend this combination of simple series to any number of differences between the two parental plants, I have indeed entered the rational domain. This seems permissible, however, because I have proved by previous experiments that the development of a pair of differing traits proceeds independently of any other differences. (qtd. in Stern and Sherwood 63)

In his reply, Mendel argues that because he has shown that the same pattern, describable with the same mathematical expression, can be found in independent empirical observations across at least three different traits, he should be able to assume that it will be the same for all traits through subsequent generations.

By investigating Nägeli's response to Mendel and Mendel's defense, it is possible to understand, at least in Nägeli's case, why Mendel's work was not compelling. To some degree it failed to be persuasive because Mendel supported Kölreuter and Gärtner's theories that the traits of species remained inalterable. This position was in conflict with Nägeli's beliefs that hybridization resulted in a completely new character as a consequence of the blending of the parents' essence during the process of hybridization.

In addition to disagreeing on substantive issues, Nägeli also took exception to the scope which Mendel claimed for his conclusions. His objections reveal that he believed Mendel had not experimented with a sufficiently broad range of organisms to make sweeping claims about heredity. Mendel's response to these objections suggests that, despite the dissonance between the scope of his claims and the scope of his observations, he still felt confident that his ability to mathematically predict outcomes across different cases offered sufficiently robust evidence to support his conclusions.

Conclusion

By investigating the ontological and epistemological commitments of Mendel and his audience, the effect of those commitments on the reception of the text, and the role of mathematical arguments in inventing and supporting his case for heredity, this chapter has endeavored to reveal some of the rhetorical dimensions of mathematical argument. A close investigation of the ontological and epistemological commitments of Mendel and his audience suggests that arguing mathematically in science can be a complex affair in which mathematical arguments, no matter how compelling or analytically robust, compete with a host of other beliefs and values which may affect whether audiences accept or reject them as legitimate warrants from which to draw conclusions about nature. Whereas Mendel, based on his experiences in the physical sciences and the results of his experimental and mathematical efforts, was compelled to believe the scope and legitimacy of his conclusions, his audiences, who were invested in oppositional models of heredity, were not.

In addition, this chapter reveals, with the help of close textual analysis, that Mendel's mathematical argument participates in processes and forms of probable argument identified with rhetoric. Careful scrutiny of the experimental design of Mendel's *Pisum* experiments and his conclusions about heredity show that Mendel relied on analogy as a source of invention for making his arguments. Examination of his experimental design suggests that mathematics provides the *phoros* in this analogy through which the *theme* of heredity could be comprehended. It also shows that, though Mendel relies on analogy as a source of invention, he tries to erase all traces of his reliance on the figure. He attempts this by moving from empirical evidence to rational principles and by dissolving the asymmetry of the *theme* and *phoros* by repeatedly arriving at the same mathematical conclusions, and then deductively deriving the same conclusions from established mathematical principles.

Arguments that Mendel was "misunderstood" or "ahead of his time" assume some *a priori* quality of truth in his work that went unrecognized by his intellectually less superior peers. An investigation of the rhetorical dimensions of Mendel's text, its context, and its reception, however, suggests that Mendel's fall into obscurity was most likely not because his audience was incompetent, but rather because Mendel's conclusions contradicted many of the established commitments of his contemporaries, and did so on the evidence from the crossing of seven traits in a single species. From this perspective, which respects the importance of collective commitments in the development of scientific knowledge, it would have been a greater shock had Mendel's audience embraced his findings.

As it turned out, changes in audience attitudes and interests as well as a vigorous campaign of argument were required before biological researchers collectively accepted Mendel's proposed connection between the rational and the empirical. With the advent of mutation theory and an ensuing priority dispute between Hugo de Vries, Carl Correns, and Erich von Tschermak in 1900 over who was the first to discover the 3:1 and 1:2:1 ratios, Mendel's largely forgotten paper was lifted back into the public consciousness. Once in their consciousness, it required the rhetorical efforts of the English biologist William Bateson, who defended the existence of pure breeding characters and petitioned well-endowed organizations such as the Royal Society to fund further research, before Mendel's conclusions were accepted and acclaimed as an important scientific breakthrough.

5 Probable Cause: Rhetorical Strategies and Francis Galton's Arguments for a Mathematical Model of Inheritance

> *It must always be bourn in mind that we are dealing with human workers who have their own ideas which must be reflected and humored if we are to gain their cordial consideration. We have, to speak rather grandly, statesmanship problems to deal with. . . . Just now, we must busy ourselves in finding out lines of least-resistance in pushing forward our nascent work.*
>
> —Francis Galton to Karl Pearson

Although it may seem historically that Gregor Mendel's experiments on sweet peas were a solitary effort to design, interpret, and argue for experimental outcomes based on the tenets of probability, the reality is that he was not the only researcher in the latter half of the nineteenth century applying mathematical principles to the investigation of inheritance. In fact, a decade before Mendel's arguments on the subject were widely read and appreciated in England, Sir Francis Galton (1822–1911) broke ground for a probabilistic/statistical approach to the investigation of heredity in his book, *Natural Inheritance* (1889).

Like Mendel, Galton developed a theory of inheritance based on an analogy between mathematical probability the hereditary process. However, unlike Mendel, Galton succeeded in persuading others to accept the analogy. But what accounted for Galton's success in persuading biologists to accept a mathematical model for heredity and variation? One of the answers is rhetoric.

Whereas Mendel adheres to conventional standards of argument in the physical sciences, moving from empirical evidence to deductive conclusions in "Experiments in Plant Hybridization," Galton adopts a more rhetorical approach in *Natural Inheritance*, taking great pains to prepare his audience to accept his conclusions and methods before introducing his data and evidence. By examining Galton's arguments, as well as the audience to which and context in which he made them, this chapter advances the conclusions that: (1) using mathematical formulae and concepts to make a scientific argument requires persuasion in cases where analogies between natural and rational patterns are not already commonly accepted, and (2) rhetorical strategies can play a substantive role in persuading scientific audiences to accept the use of mathematical formulae and concepts to make claims about nature.

Method

In order to establish the impact of rhetorical strategies on the acceptance of the mathematical analogy in Galton's *Natural Inheritance,* this chapter adopts a three-phase method of analysis. To understand the influence of extra-textual circumstances that may have shaped the choices in text or conditioned its reception, the first phase of analysis will examine (1) Galton's *situated ethos*, or reputation, within the scientific community prior to the publication of the text, (2) the audience's level of understanding and acceptance of the mathematical/biological analogy before the creation of the discourse, and (3) the persuasive efforts prior to *Natural Inheritance* made by Galton and others to support the analogy between mathematical probability, heredity, and variation.

Whereas the first phase of analysis attempts to paint a picture of the intellectual and rhetorical context out of which the text emerged, the second phase focuses on the text itself to illuminate the strategies Galton uses to make his case that hereditary outcomes and the distribution of values calculated from the law of error were analogous. This textual analysis provides evidence of Galton's dependence on rhetorical methods, particularly the use of narrative and appeals to the values of his Victorian audience to persuade them of the utility of his mathematical approach.

The third and final phase of analysis examines the response of Galton's audience to the text and the general influence it had on the development of the subdiscipline of mathematical biology, known as biometrics, in the last decade of the nineteenth century. Analysis of the reviews of *Natural Inheritance* offers evidence not only that readers of the text found it compelling, but also that particular aspects of the text, including the rhetorical strategies, contributed meaningfully to its acceptance.

Historical Context: Situated Ethos, Audience, and Prior Argument

Ethos: Francis Galton, Gentleman of Science

In many ways, the context to which Galton responded as he crafted his text was not significantly different from the one Mendel was facing. Both men had an uphill battle to fight. Their audiences of biological researchers were likely more knowledgeable about biology than mathematics, and generally skeptical about there being an analogy between probabilistic outcomes and hereditary ones.

In one respect, however, the two men couldn't have been more different. Whereas Mendel was a generally obscure scientific figure toiling away on his experiments in Brünn far away from the intellectual circles of Vienna and Berlin, Galton was a gentleman of science in the thick of London intellectual society. This section explores Galton's *situated ethos*, and makes the case—supported later by evidence from the audience response to *Natural Inheritance*—that the importance of its influence on Galton's success with his audience should not be underestimated.

Science and mathematics have often been considered non-rhetorical because of the degree to which they attempt to eliminate and, in fact, challenge the importance of ethos in establishing belief or knowledge. This particular perception has been widely challenged by rhetoricians of science. They have illuminated how judgments about the credibility of scientific conclusions often depend on the stature of a scientific arguer or his/her sources, and how scientific arguers construct an ethos for themselves in their writing by employing disciplinary-specific methods, language, etc.[1]

A dimension of ethos in scientific writing that has not been as widely explored, however, is how changes in the level of certainty of scientific conclusions raise or lower the importance of ethos as a persuasive dimension of scientific argument. Such a link is made by Carolyn Miller in her article, "The Presumption of Expertise: The Role of Ethos in Risk Analysis," in which she reveals that when traditional probability is replaced by a less certain, subjective probability, the ethos of scientific experts who do the guessing becomes an important warrant for supporting mathematical conclusions (Miller 175–85).[2] Although Miller's interest is ultimately in the attempts to mask the importance of ethos in scientific argument, the evidence she presents of the ascendency of ethos in the absence of empirical evidence makes a strong case for a correlation between the degree of certainty of knowledge and the importance of ethos.

In the case of Francis Galton's argument, ethos plays an important supporting role because his conclusions are largely uncertain. Evidence presented in the next two sections about the extra-textual context reveals that there were a number of grounds on which his arguments would have been considered unreliable or illegitimate by his audience. For example, there was no general agreement in his audience that human heredity was sufficiently comparable to the cases in the physical sciences to which the law of error had traditionally been applied. In addition, the majority of his intended audience did not understand the mathematics he was using.

Though Galton does his scientific due diligence, drawing conclusions on the basis of quantitative empirical evidence connected with mathematical warrants, without substantial credibility the text might have languished in obscurity, as Mendel's work initially had, with little impact on the direction of the scientific community. Galton's fate, however, was different because he was a well-known and well-regarded member of the nineteenth century British scientific aristocracy.

By the end of the nineteenth century, Francis Tertius Galton represented one of the last of what was becoming a quickly extinct species: the gentleman scientist; a man with the financial freedom to bankroll his own research and social connections to have the results of his work given a serious hearing. From birth, Galton enjoyed a comfortable position within the highest intellectual and social strata of English middle-class society. His father, Samuel Tertius Galton (1783–1844), was a successful Quaker industrialist, and his mother, Violetta (Darwin) Galton, was the daughter of Erasmus Darwin and aunt to Charles Darwin (Brookes xv, 4). His father Samuel was an amateur scientist who contributed to the theory of color vision. His grandfather, Erasmus

Darwin, was a famous member of the intellectual society the Lunar Club who devised a steering mechanism for carriages, published a classic paper explaining the formation of clouds, and crafted the poem "Zoonomia," that argued that all living things were descended from a single, microscopic ancestor (Bulmer 3).

Galton made his first mark as a gentleman of science by funding an expedition to map previously uncharted territory in the dark heart of Africa. Originally, the expedition began as a vacation; however, through a stroke of fortune, it became an expedition which put Galton on the path towards a life-long pursuit of scientific knowledge (*Memories* 122–23). The inspiration to make the trip a journey for science initially came from Francis's cousin, Douglas Galton, a Fellow of the Royal Geographical Society, who suggested that Francis might incorporate in his adventure some scientific activity. His cousin provided him with introductions to some of the Geographical Society's members with whom he develop a plan of exploration for his journey that was to take him from Cape Town to Lake Ngami, recently discovered by David Livingstone.

Members of the Society suggested that Galton survey the land on his route and report back his findings (Bulmer 11). Although unversed in the use of the sextant and techniques of surveying, through reading and practicing Galton became competent enough to use the basic tools and techniques of geodesy to survey the country through which he traveled, making careful measurements of longitude and latitude (*Memories* 125–26). He sent his observations back to the Royal Geographical Society, who published them in their journal in 1852. In 1854, because of his work in Africa, Galton received one of the society's two annually awarded gold medals, "for having at his own expense and in furtherance of the expressed desire of the Society, fitted out an expedition to explore the center of South Africa, and for having so successfully conducted it . . . as to have enabled this Society to publish a valuable memoir and map in the last volume of the journal" (*Memories* 150).

Galton's success in his African adventures helped him establish a credible reputation in the scientific community and encouraged him to devote his time, energy, and wealth to the pursuit of science. In his memoirs, Galton writes of the ethos that his honors from the society afforded him: "The Geographical Medal gave me an established position in the scientific world. In connection with subsequent work, it

caused me to be elected a fellow of the Royal Society in 1856, and to receive the very high honor of election to the Athenaeum Club" (151). With these new laurels and a disposable income to help fund his own scientific research, Galton was soon an established member of the scientific elite. He was elected to the council of the Royal Geographical Society in 1854, and made secretary in 1857. He was also elected to various high positions in the British Association for the Advancement of Science, including general secretary (1863–67), president of the Geographical section (1862, 1872), and president of the Anthropological section (1877, 1885), turning down the position of President of the entire society twice (Gillham 105).

Common and Special Warrants

In the first fifteen years of his scientific career, Galton devoted himself primarily to geodesy, the study and measurement of the shape of the earth, and the study of the weather. These were areas of investigation in which the work was primarily conducted by making precise measurements and using abstract mathematical formulae to describe geodesic and meteorological phenomena. By the middle of the 1860's, however, Galton's scientific work began to shift, subtly at first, to questions about heredity inspired by the publication of his cousin Darwin's landmark work *The Origin of Species.*

The Origin of the Species provided Galton with a conceptual framework for assessing the variation of natural populations statistically and, importantly, focused his attention on heredity as a phenomenon central to understanding natural variation. What Galton hoped to contribute to the work started by his cousin was a robust mathematical theory of heredity that would describe and, thereby, explain the distribution of traits from one generation to the next. In this turning towards the murky pre-paradigmatic science of heredity and the radical application of mathematical probability to its study, Galton's reputation, both earned and inherited, became an important pillar of intellectual support. This support would become necessary because the analogy between biological phenomena and the mathematical law of error was not generally accepted by his audience.

To comprehend why the analogy posed such a problem, it is necessary to understand the differences between the *epistemic commitments* of the physical and biological sciences. The concept of epistemic commitment is an expansion on Karin Knorr-Cetina's idea of *epistemic cultures*, which she defines as "cultures that create and warrant knowledge" (1). For Knorr-Cetina, epistemic cultures can be represented by their "machineries of knowledge construction," an integrated fabric of instrumental, linguistic, theoretical, and organizational frameworks (10). The conceptual expansion being proposed here is to use the phrase "epistemic commitments" to convey the totality of shared commitments that constitute the common ground of ideas, values, beliefs, and language that Perelman and Olbrechts-Tyteca explain are the starting point for argument (65). Though Knorr-Cetina talks about linguistic and theoretical frameworks within an epistemic culture, these frameworks are not considered aspects of argument. The phrase "epistemic commitments" is intended to extend the notion of epistemic culture to include this dimension.

The epistemic commitments in a particular epistemic culture or group of cultures can help make sense of why arguments that cross borders, and/or represent novel forms of argumentation, succeed or fail. These commitments provide benchmarks from which aspects of a particular argument can be judged to fit or not to fit within the accepted framework of an epistemic culture. In most cases, however, it is not a whole system of epistemic commitments that are involved, but specific commitments or *warrants*, "hypothetical statements which can act as bridges, and authorize the sort of step to which our particular argument commits us," that are at stake (Toulmin, *Uses of Argument* 91).

As individual aspects of the totality of epistemic commitments of an epistemic culture, warrants, as a category, can include a broad range of constituents. They embrace, for example, lines of argument that have endured the test of time to become standard strategies for reasoning (*topoi*), such as cause and analogy. However, they also include principles, or "rules of thumb," that are officially or unofficially established as knowledge within the epistemic culture (Toulmin, *Introduction to Reasoning* 213–214). Though the comprehensive valence of the term "warrant" makes it amenable to discussions of scientific argument, it also creates a challenge in cases, such as this one, in which distinctions need to be made between a warrant that is commonly accepted for argument, and one that has limited or no acceptability. In order to mark these distinctions, the category *warrant* will be further divided into *common* and *special* warrants. The phrase *common warrant* will be used to designate warrants that are neither challenged nor preemptively defended in an argument, and *special warrants* to refer to authorizing principles or lines of argument that are challenged when used, or require preemptive defense.

Special and common warrants can be used as tool for designating the parameters of epistemic commitments in epistemic cultures because they illuminate the principles, or lines of arguments, that members of a community accept and count as part of their repertoire, as well as what principles or lines of arguments they reject. As we shall see, the special status of a warrant can be the consequence of a number of factors, including its novelty and its incompatibility with already accepted principles or lines of argument. Because of their shortcomings, special warrants require the support of further data, common warrants, and rhetorical strategies—including appeals to the beliefs and values of the imagined, universal audience or the ethos of the speaker—to win assent.

The remainder of this section will be devoted to making the case that while there were some common mathematical warrants that were acceptable epistemic commitments in the culture of the non-physical sciences of the nineteenth century, these commitments did not include the mathematical law of error. To make the case that Galton's arguments for heredity relied on special warrants for support, we will first identify the warrants, and then make the case that, (1) most researchers studying populations of organisms rejected them, and (2) those researchers who did accept them were forced to preemptively defend them against criticisms they knew would arise in their audience. Establishing the special status of the warrants supporting Galton's mathematical arguments about variation and heredity provides the necessary context for understanding the importance and necessity of Galton's use of rhetorical strategies for argument in *Natural Inheritance.*

Common Warrants. In *The Origin of the Species,* Darwin's use of quantification and basic arithmetic both to create an ethos of precision and rigor, and to invent and support his arguments, represented well-worn, commonplace strategies for describing nature, and had precedence in the work of Candolle, Humboldt, Brown, and other biogeographers. Most natural philosophers accepted that types could be counted—though they might disagree about whether an organism should be categorized as one type or another—and they believed that comparisons of greater or lesser could be made between groups of organisms using ratios. The commonality of these warrants is supported by the fact that none of these methods for argument was challenged or rejected by Darwin's audience. As a consequence, the described and calculated values Darwin uses are all accepted as common warrants, or certified positions of strength, for making his arguments.

The same cannot be said of Galton's use of the mathematical *law of error* to establish conclusions about the hereditary processes and outcomes. Whereas Darwin mostly started with collected data and then used quantification and arithmetical comparisons to make standard arguments for his position, Galton, who had experience in applying mathematical formulae to meteorological and geodesic data, began his investigation of heredity, like Mendel, with mathematical concepts and principles from the physical sciences as hypotheses, and then set out to apply them to heredity through empirical investigation. The sophistication of the mathematics and their unconventionality in the non-physical sciences had important implications for Galton's argument.

Victorian Statistics and Mathematical Probability. In order to understand what the *law of error* is and why Galton's arguments from it were considered special warrants by late nineteenth century social and biological researchers, it is necessary to understand the law's original development and parameters for use in the physical sciences. Investigating these historical facets helps identify the warrants underlying the law and explain exactly why it inspired resistance.

The mathematical *law of error* was developed by astronomers out of a need to identify the "true value" of a particular observation. From the early period of the European astronomical tradition in the seventeenth century, it was clear to astronomers measuring celestial objects night after night that, no matter how carefully they measured, the results they obtained from one evening to the next were always slightly different. These slight differences, they assumed, were not due to causes that had primary influence on the objects they were measuring, but was rather associated with accidents or errors, a multitude of minute causes collectively unknown and unknowable for the measurer.

In the earliest stages of preoccupation with the true value of measurement, errors were accounted for by averaging the outcomes of observations. Despite the simplicity of the method, some astronomers were still haunted by the question, "How do I know that the average number is the true value of the observation?" One of the most important conceptual advances in answering this question was made by Pierre-Simone Laplace, who argued in his *central limit theorem* that there was an equal probability of a measurement going over or under the true value of a measured effect. He also suggests that, because big mistakes are rarer than small mistakes in observation, it is more probable that a given measurement is closer to the true value of an observed phenomenon than farther away from it. All of these parameters are illustrated in the bell curve, where the actual value is located at the highest point in the curve, and other observational values affected by error fall in frequency in equal measure from this maximum value down both sides of the curve (Daston 271).

Though Laplace's theorem introduced some important assumptions about error into mathematical thinking, it was not sufficiently robust to offer a reliable calculation for the distribution of error. The honor for calculating this distribution goes to one of Laplace's contemporaries, the German mathematician/astronomer Friedrich Gauss (1777–1855). Whereas Laplace began with models of error and then tried to fit them to models of observational results, Gauss began by assuming theoretically that the mean of an observational sample is a good estimate of "true value." Starting from this assumption, Gauss then considered what form of density function coincided with the mean in the observational sample (Chatterjee 226). The answer, based on this method of calculation, was the normal distribution described in the eighteenth century by the mathematician Abraham de Moivre in, *Miscellanea Analytica* (1730) and *The Doctrine of Chances* (1718, 1738, and 1756).

Gauss's development of the law of error represented a watershed in mathematical probability and astronomy because, for the first time, a mathematical method existed whereby the "true value" of a set of observations could be mathematically substantiated by comparing it to a predetermined distribution of errors. This index of verity could, on the one hand, be a powerful tool for identifying fixed patterns in nature and, on the other, the true value of any measurable phenomena. These capabilities made it attractive to astronomers and geographers in their study of the heavens and earth; however, they also supported a greater potential for general philosophical interest in the possibility of demonstrating other underlying patterns in nature.

While physical scientists enthusiastically embraced Gauss's mathematical method of ascertaining the "true value" of a phenomenon, a similar enthusiasm for an expanded role for mathematical probability in the assessment of statistical data was not commonly shared by mid-Victorian statisticians. As historians of mathematics and science point out, the majority of researchers using statistical methods to study social and biological phenomena, unlike their counterparts in astronomy and geodesy, did not employ mathematical probability in their work. This absence was due in part to a general lack of exposure amongst researchers in these fields to advanced mathematics, and more importantly, to a widely shared belief that, unlike physical phenomena, biological, social, and psychological phenomena were not amenable to study using mathematical probability.

A lack of exposure to advanced mathematics was a serious impediment to developing mathematical applications from the law of error. The dearth of higher mathematical experience is suggested by the types of occupations in which members of early statistical societies were engaged. In the London Statistical Society, for example, there were, according to historian Philip Abrams, "a few mathematicians among the original members; however, there were many more economists, politicians, peers, government officials, and doctors of medicine" (14). Though this learned constituency was not without the general skills requisite for basic mathematical manipulations, it is unlikely that any but those members whose primary vocation led them to study continental mathematics, particularly French or German mathematics, would have been exposed to mathematical probability or its basic principles (Abrams 14, Mackenzie 9).

In addition to the lack of knowledge of advanced mathematics, the development of statistics informed by mathematical probability was inhibited by the belief amongst Victorian statisticians that mathematical probability could not be applied to statistical aggregates of biological phenomena. As it was originally conceived, the probabilistic law of error was developed in astronomy and geodesy rested on two important assumptions: (1) that in each instance, the *same phenomenon* was observed and 2) that in each instance, the *same causes* were underlying the effect being measured. These two conditions are intimately intertwined. A unique phenomenon such as the motion of a single planet, for example, was assumed to be acted on by generally the same causes each time it was measured. As a consequence, all measurements were sufficiently homogeneous to be comparable. Therefore, an astronomer could assume that any variations in the value of their measurements was either the result of irregular error caused by a collection of numerous, imperceptible causes, or the result of regular error from weather conditions or instrumental malfunction.

When assessing social and biological phenomena, however, there was not the same confidence in the homogeneity of observations. First, there was the most obvious problem that, instead of observing the same planet or points on the earth over and over many times, statistical investigations measured many unique instances a single time. Second, and perhaps most important, was the problem of causation. Statistical research was concerned with aggregates of individuals living in varying conditions, across diverse locations. Because there was no clear understanding of the impact of the environment or behavior on human conditions, there could be no certainty amongst statisticians that their sample populations were sufficiently similar to be assessed with the law of error.

Instead of focusing on expanding the applications and/or mathematical principles of probability, which assumed *a priori* regularity in the phenomena being measured, much of the earliest work in statistical research was aimed at trying to establish causal relationships between different factors such as, poverty and morality, living conditions, and illness. As Stephen Stigler explains in the *History of Statistics:*

> Social scientists . . . for much of the [nineteenth] century . . . groped toward some acceptable way of overcoming the inherent diversity of their material. They tried to overcome the conceptual barrier to combination of observations with brute force, with massive data bases that they hoped would inform them of all important causal groupings. But the result was frustration, even among those pioneers who had made some progress. (5)

Though probability eventually played an important role in assessing the relationship of different situational factors to one another and to the cause being investigated, the conceptual framework for developing these applications in the social and biological sciences was not firmly in place amongst nineteenth century statisticians or mathematicians interested in probability. [3]

A brief historical overview of the development of the law of error and its challenges for statistical researchers studying biological phenomena suggests that it was considered within these epistemic cultures not to be a facet of their epistemic commitments. Further, historical evidence suggests plausible reasons for its exclusion. On the one hand, the lack of training in advanced mathematics amongst social and biological researchers gathering statistics likely made the application of the law of error alien to their regular practices of argument in the discipline. On the other hand, there was a general philosophical resistance towards the central warranting assumptions for making an analogy between the law of error and organic populations, which raised questions about whether biological and physical phenomena shared sufficiently similar properties to be usefully assessed using this mathematical method.

Rhetorical Preludes to *Natural Inheritance*

In addition to examining historical evidence about the development of the law of error and its use among statistical researchers, the law's status as part of the epistemic commitments in the biological sciences can also be established by examining the arguments of researchers who did use it to make claims about variation and heredity in organic populations. If we accept that special warrants are "special" on the grounds that they require a defense when used as arguments within a particular epistemic culture, then proof that they required defense would support their characterization as special warrants. A brief examination of Galton's first text employing the law of error to draw conclusions about heredity, *Hereditary Genius*, reveals that even before *Natural Inheritance*, Galton recognized that his warranting assumptions would face resistance even though he did not fully appreciate the degree of effort and the importance of rhetorical strategies in overcoming this opposition.

In *Hereditary Genius*, Galton argues that the law of errors accurately describes the distribution of variation in the traits of human populations. In the third chapter of the book, entitled, "Classification of Men According to their Natural Gifts," Galton applies the *law of deviation from the average* (i.e., the law of error) to a collection of heterogeneous subjects to make his case about the distribution of eminence within a population:

> I propose in this chapter to range men according to their natural abilities, putting them into classes separating them by equal degrees of merit, and to show the relative number of individuals included in the several classes. . . . The method I shall employ for discovering all this, is an application of the very curious theoretical law of deviation from the average. (26)

As Galton develops his arguments in the chapter, he is aware of the complications associated with judging phenomena that are at once biological, psychological, and sociological, as similar to physical phenomena. In order to defend his analogy, and thereby his method, he makes efforts to assure readers that his human subjects are comparable to physical phenomena. He accomplished this by explaining that (1) the subjects he has chosen come from a sufficiently random sample and, as such, cannot be influenced by causes other than the multitude of minute, random causes that are associated with probability and (2) outcomes are not significantly influenced by other causal factors, like nurture, which would seriously skew his distribution and thereby challenge the legitimacy of what he might claim from his data.

As an introduction to the mathematical method in chapter two of *Hereditary Genius*, he tantalizes the reader with some shocking results about the intellectual prowess of students in mathematics at Cambridge. Before he provides these results and discusses his mathematical method, however, he takes measures to persuade his readers of their validity by assuring them that his sample was sufficiently random. At the beginning of his presentation, he asserts, "No doubt the bulk of Cambridge men are taken at hap-hazard. A boy is intended by his parents for some profession . . . he should be sent to Cambridge or Oxford" (17). In this passage Galton suggests that, like two sides of a coin flipped at random, all boys either end up Cambridge or Oxford. This representation of the school selection process supports Galton's analogy between student intellect and the law of error by suggesting that the process underlying that selection has all the hallmarks of a random occurrence. As a consequence, it can be judged using the same mathematical tools for describing the outcome of probabilistic events.

In addition to making the case that the underlying school selection process is probabilistic, Galton also establishes that there are no primary causes—particularly class and nurture—that might significantly affect the distribution of scores, and thereby his conclusions, about the distribution of intelligence in Cambridge students. To dispense with any critique that there might be about other mitigating factors affecting intellectual outcomes, Galton begins the chapter by strongly asserting that, despite popular views to the contrary, there are no non-heritable conditions that might affect the absolute aptitude of a human being:

> I have no patience with the hypothesis occasionally expressed . . . that babies are born pretty much alike, and that the sole agencies in creating differences between boy and boy, man and man, are steady application and moral effort. It is in the most unqualified manner that I object to pretentions of natural equality. (14)

With these sentiments, Galton hopes to dispatch the great influence that personal effort and socio-economic factors might have on intellectual outcomes. However, to ensure that, for his particular subjects (students of mathematics at Cambridge), there is no room to doubt there is an equal playing field, he adds:

> The youths start in their three year race as fairly as possible. They are then stimulated to run on the most powerful inducements . . . and at the end of the three years they are examined most rigorously according to a system that they all understand and are equally well prepared for. (18)

Galton's argument for the stochasticity of his sample, and his dismissal of all possible primary causes that might have a substantive effect on the phenomena he is measuring, is an effort on his part to warrant his use of the law of error. Galton assures his audience that, like physical phenomena, psycho-biological phenomena can be considered sufficiently homogeneous and probabilistic to be reliably described by the law of error. Despite these efforts, the case for his analogy between the law of error and heredity was not sufficiently robust to win the general assent of his audience. In one, anonymous review of *Hereditary Genius* in *Galaxy* (an illustrated magazine of entertaining reading), the reviewer writes:

> We doubt whether Mr. Galton's ingenious and elaborate lists and tables really carry us much beyond the point from which we may see that there is some substantial reason for admitting the aptness and probability of the theory. . . . Our own

> impressions are with Mr. Galton rather than against him . . . but he has hardly pushed his theory high enough up into the clear dry air of scientific debate to secure it as yet a chance of full, grave, and exhaustive consideration. (424)

In these lines, the reviewer from the *Galaxy* voices his skepticism about whether Galton's mathematical methods were truly delivering what they promised. This raises the question: Why weren't Galton's lists, tables, and other evidence compelling for the reader?

Another review of the book in the *Journal of Anthropology* offers some answers. In the review, author George Harris—a judge and amateur psychologist/anthropologist—makes a series of suggestions to Galton about future directions he should take in his work. All of these suggestions relate to causes such as psychological responses to parents and social position, which Galton tries to dismiss as unimportant but which the reviewer clearly believes are mitigating factors that Galton needs to consider. He remarks about the effect of social status on political genius, for example:

> In the case of legislators and statesmen, the fact of the son following in the same occupation as the father, and with a certain amount of success, can hardly be allowed to be a proof of hereditary genius. . . . His rank and position naturally induce him to take part in public life, and he acquires at once a prominence, not from his abilities, but from his standing and fortune. (61)

These negative reviews suggest that Galton had not made a sufficiently compelling case for the analogy between the law of error and probability. In part this was because he had not been effective in establishing the warranting assumptions that made the analogy feasible.

Although there were critiques of Galton's mathematical methods and warrants, there were also words of support from reviewers who applauded Galton for attempting to develop a more precise approach to the study of heredity. In the *Atlantic Monthly*, for example, an anonymous reviewer writes:

> There is a great deal of loose thinking current, both as to the kind and to the degree of innate differences of capacity between different men, and as to the mode in which such differences are transmitted from parents to children. Upon both of

> these points Mr. Galton furnishes ingeniously arranged data for forming precise estimates. (753–54)

Perhaps more significant than a general approval of Galton's mathematization of heredity, however, is the reviewer's seeming acceptance of at least one Galton's warranting assumptions: that the students of mathematics are sufficiently homogenous and unaffected by outside, major causes to be comparable. Following Galton's lead, he advises the reader: "Let us remember that they [the students] are all working to the utmost limit of their capacity . . . and that, in general, they have had about equally good opportunities for preparation" (754).

A similar appreciation for the precision of Galton's mathematical argument can be found in other positive reviews of the book; however, there is no further evidence that Galton's warrants had been recognized and adopted by these reviewers. In the *Westminster Review*, for example, a somewhat skeptical but still supportive anonymous reviewer characterizes Galton's work as an important extension of mathematics to the study of heredity, but has some reservations about the veracity of its conclusions:

> The first to treat the topic [of heredity] in a statistical manner, or to arrive at numerical results, and to introduce the law of deviation from the average into discussions on heredity, Mr. Galton has a special claim on the thoughtful attention of his readers, even though his book should be found occasionally inaccurate or deficient. (144)

A brief review of audience responses to *Hereditary Genius* suggests that Galton's efforts to establish an analogy between the mathematical law of error and heredity had a mixed reception. On the one hand, there was support for the book because it provided readers with, to their knowledge, the first mathematical treatment of heredity, which was a valuable undertaking because it made the study of heredity a more precise endeavor. On the other hand, there was a degree of skepticism about the conclusions and the underlying warrants that supported them. Critics suspicious of the robustness of the book's conclusions challenged the simplicity of Galton's assumptions, and thereby the ability of statistics and probability to say something meaningful about a subject as causally complex as heredity.

In the two decades between the publication of *Hereditary Genius* (1869) and *Natural Inheritance* (1889), Galton worked assiduously to strengthen his case, gathering new evidence to support the mathematical analogy and developing more sophisticated mathematical tools and arguments to explore its implications. Drawing from his personal fortune, Galton began the herculean task of collecting his own data. In 1884 he sent out a notice offering a reward for qualitative and quantitative data from family records for such characteristics as stature, eye-color, temper, and artistic faculty (Galton, *Natural Inheritance* 72, 77). In the spring of the same year, he also set up an anthropometric laboratory at the International Health Exhibition in the Gardens of the Royal Horticultural Society, where visitors paid three pence to have their vital statistics recorded, including, eyesight, hearing, height, and strength (Gillham 211–12). From these statistics, and statistics from a more permanent anthropometric laboratory set up in the Scientific Galleries of the South Kensington Museum, Galton was able to obtain information on more than seventeen different characteristics for over 13,000 individuals.

Using these statistics, Galton prepared a series of papers that were accepted for publication in both intra- and inter-specialist journals in 1886, including: *Nature*, *Proceedings of the Royal Society*, and the *Journal of the Anthropological Institute.* These papers offer a variety of conclusions from his data, including the confirmation of his mathematically derived law of ancestral heredity,[4] and the novel claim that the value of a particular hereditary trait regressed towards the mean as it moved from one generation to the next.[5]

The number of reputable journals that accepted Galton's papers suggests that, despite previous skepticism, his massive collection and mathematical analysis of data was sparking some general interest in the application of probability to the study of biological phenomena amongst natural researchers. This potential growth of enthusiasm provides an important context for understanding the exigency for *Natural Inheritance.* With his ideas in ascendance, the book afforded Galton the means to codify his intellectual position and attempt once more to attract a following for his ideas.

A brief investigation of the historical context preceding the publication of Galton's work *Natural Inheritance* reveals that Galton had accumulated a lot of credibility as a scientist and a great deal of experience as a researcher of heredity, but faced an audience who was skeptical about the applicability of the law of error to the study of biological phenomena. Galton's personal history suggests that he was born into the intellectual aristocracy of Victorian England, but had made concerted efforts to earn his laurels as a respected researcher in the physical sciences. Although these aspects of his *situated ethos* were not sufficient to impact the reception of *Hereditary Genius*, his later, more substantial efforts to gather data and further develop his theory of heredity, in conjunction with previously established ethos, likely influenced the reception of *Natural Inheritance.*

In addition to providing detailed information about Galton's ethos as a scientist, examining the historical context in which *Natural Inheritance* was developed also provides valuable information about the epistemic commitments of the non-physical sciences, and thereby the challenges that Galton faced in making his case for a mathematical analogy between the law of error and heredity. This analysis reveals that Galton faced resistance from his audience because they did not accept the fundamental warrants necessary to support an analogy between the law of error and heredity. Although he made some effort to recognize and address his audience's skepticism in *Hereditary Genius*, this effort was only partially successful.

Natural Inheritance: The Case for a Mathematical Science of Heredity

Part two of this chapter examines Galton's second major effort at making his case for an analogy between heredity and the law of error, *Natural Inheritance* (1889). Unlike his first attempt, Galton's second effort succeeded in persuading his audience to accept his warrants and inspired at least some of them to begin developing new approaches of analyzing nature based on his methods. The question that this part seeks to answer is: Why?

Although evidence has already been presented that there was a softening in the disposition in his audience of anthropologists towards his statistical approach based on the new data he collected, I would also contend that his eventual success in making his case was aided by choices he made about *what kind of* information to present and *how* to present it in *Natural Inheritance.* The close reading of the text that follows suggests that Galton believed that a variety of strategies were necessary to convince his audience. In some cases, these strategies are logical and empirical, adhering to the conventions of formularization, verification, and extrapolation described by Whewell and Herschel in their discussions of the conventions of scientific argument. In other cases, however, they are rhetorical, employing narrative and appeals to the values of the audience to affect persuasion.

Close textual analysis of the argument the first five chapters of *Natural Inheritance* support these claims. They reveal that Galton moves from rhetorical to empirical, empirical to logical, and logical to deductive claims with each level of argument, depending on the previous one for support. This careful building of argument suggests that, because of past skepticism towards his warrants, and therefore, his methods, Galton believed he needed to employ extensive rhetorical strategies to establish their validity. Only then, does he turn to his data and mathematical reasoning to empirically and analytically establish his conclusions about heredity.

Chapters Two and Three: A Hereditary Tale

At the very beginning of the book, Galton makes his case rhetorically, starting with the establishment of the warrants supporting his analogy between the mathematical law of error and heredity. In chapters two and three, "Process in Heredity" and "Organic Stability," he devotes extensive attention to the task of crafting a compelling explanation of the biological processes of heredity that support his contention that biological processes are probabilistic and that biological traits are homogeneous. To support and explain these processes, he turns to the rhetorical strategy of narrative.

Narrative has long been considered an important tool for argument in rhetorical theory, both as a strategy for intensifying an experience and as a tactic for reasoning about events. In forensic argument, for example, narratives play a key role in establishing the facts of a case, that such and such a thing did or did not, could or could not, have happened. Aristotle recognizes the role of narrative in forensic oratory in *Rhetoric* when he offers advice on how to structure a forensic narration. He explains that correctness in making a narrative lies "in saying just so much as will make the facts plain, or will lead the hearer to believe that the thing has happened, or that the man has caused injury or wrong to someone, or that the facts are really as important as you wish them to be thought: or the opposite facts to establish the opposite arguments" (III. 16 1417a).

Existing rhetoric of science literature clearly recognizes the connections between forensic and scientific argument. In Jeanne Fahnestock's groundbreaking article "Accommodating Science," for example, Fahnestock categorizes scientific arguments as forensic on the grounds that, like legal arguments, science deals with questions of fact and discusses events, experiments, or observations that have happened in the past. She argues: "A case can be made for classifying original scientific reports as forensic discourse. Scientific papers are largely concerned with the validity of the observations they report" (333). Even though she admits they also contain portions of text that are deliberative, she maintains, "scientific papers are, for the most part, explicitly devoted to arguing for the occurrence of past fact" (333).

Historical investigations of scientific argument corroborate the forensic affinities of science, and suggest that, at least in its very early stages, scientific arguments had many of the features of forensic narrative argument. In *Communicating Science* Alan Gross, Joseph Harmon, and Michael Reidy explain that in seventeenth century science, a substantial amount of scientific argument was made by narrating events the natural philosopher had witnessed: "early scientific articles seek to establish credibility more by means of reliable testimony than by technical details, more by qualitative experience than by quantitative experiment and observation" (34). Like arguments in a court of law, the validity of these narratives could be considered robust if there were witnesses and if the arguer made testimony under oath: "English scientists seem to lean more heavily upon reliable testimony concerning new facts about the natural world. . . . Scientific evidence is viewed exactly as forensic evidence would be" (53).

As science developed, the features of the forensic narrative in science, such as witnesses and evidence, began to transform and evolve in new ways. Evidence accumulated and theories of physical phenomena became more robust. As a consequence of these changes, reproducible experiments and citations began to take the place of witnesses. Narrative based argument derived from commonsense and shared experience gave way to the special rationality of mathematical proof and established, observational and experimental knowledge (Gross, Harmon, and Reidy 224).

As scientists made efforts to uncover the timeless truths of nature, the very goal of their work began to run counter to traditional expectations for narrative practices and forensic arguments. Narrative's focus is commonly to describe a discrete set of events and/or actions in time, or a related set of discrete actions and or events across multiple times.

Rarely, however, is narrative involved in the habitually true, those sets of actions or processes that are achronic, or timeless. Narrative must exist in time. Likewise, forensic arguments are also seldom interested in anything but a discrete past incident or set of incidents which are the subject of legal interrogation.

Scientists, however, except those working on historically contingent events, are generally interested in developing theories which describe regular patterns of activity or behavior that are achronic. This tendency is hinted at in Latour and Woolgar's discussion of the activities of scientists in *Laboratory Life*, in which they list near the top of their value scale of statement types, "type four statements," which require no modalities when used (77). Statements such as, "Ribosomal proteins begin to bind to pre-RNA soon after its transcription starts," lack tense, aspect, or modality because they exist as facts that are true in the past, present, and future (Latour and Woolgar 77).

Although changes in scientific practices and knowledge were fundamental consequences of the scientific revolution, not all areas of natural philosophy were equally affected by them. While great strides, for example, were made in understanding physical phenomena such as the tides and the motion of the planets using these new methods, non-physical sciences like evolution and geology were only beginning to develop stable paradigms in the nineteenth century, and, as a consequence, more specialized conventions for evidence and argument. In some subdisciplines, such as the study of heredity, however, so little was known and agreed upon about the phenomenon that arguments could still employ the forensic narratives of protean seventeenth century science without their being dismissed out of hand as illegitimate.

In *Natural Inheritance*, Galton attempts to establish the special warrants necessary to make the analogy between heredity and the law of error based on a mixture of forensic and hypothetical narrative and more conventional scientific descriptions of processes. This mixed strategy for argument highlights the fundamental problem that Galton faced in making his case. Though there were commonsense beliefs about heredity as well as specialized but unproven theoretical commitments on the subject, there was no firm commitment to a single theory or sufficient empirical evidence to support a particular position. As a consequence, to make a case, an arguer had to appeal to general forms of evidence and illustration to support his point. The importance of these tactics offers compelling evidence that rhetorical strategies can play a critical role in the establishment of mathematical arguments in science.

In the opening lines of chapter two, Galton clearly explains to the reader his method and goal for argument: to present a description of the process of heredity for the purpose of justifying the mathematical method of his investigation.

> A concise account of the chief processes in heredity will be given in this chapter, partly to serve as a reminder to those to whom the works of Darwin especially, and of other writers on the subject, are not familiar, but principally for the sake of presenting them under an aspect that best justifies the methods of investigation about to be employed. (4)

As the primary witness of authority to corroborate his conclusions, Galton calls to the stand his cousin Charles Darwin, whose theory of pangenesis Galton uses as a model of heredity with some important modifications. According to Darwin's pangenesis theory, the cells that are created in the zygote contain minute particles called "gemmules," each of which represents a different physiological feature inherited from the parent. As the cells multiply, they throw off these small particles which, "when supplied with proper nutriment, multiply by self-division, and are ultimately developed into units like those from which they were originally derived" (Darwin, *Variation of Animals* 374).

Darwin's basic concept of the particulate gemmule is clearly echoed in the opening section of Galton's second chapter in which he narrates the process by which traits are inherited. Even without the concept of a gene—the term "gene" is not coined until 1911—Galton makes a case for particulate inheritance (OED).

> We seem to inherit bit by bit, this element from one progenitor that from another . . . while the several bits are themselves liable to some small change during the process of transmission. Inheritance may therefore be described as largely if not wholly "particulate," and as such it will be treated in these pages. (7)

Up to this point, Galton's explanation is a fairly standard scientific description. However, immediately following his claim for the particulate nature of heredity, Galton turns to an illustration of the process of particulate inheritance: "Though this word ["particulate"] is good

English . . . the application now made of it will be better understood through an illustration" (7–8). The example that follows likens particulate heritable traits to the building materials cast off from old buildings, which are then reused by builders to create new buildings. The consequence is a bricolage, a building which incorporates architectural features of all of the buildings used to create the structure.

At this point in the illustration, Galton turns to narrative to explain the process of heredity. He asks the reader to imagine that he and the author "were building a house from second hand materials from the dealer's yard," and asserts that, "we should often find considerable portions of the same old houses to be still grouped together" (8). After an additional few lines of vignette, Galton transitions from the hypothetical narrative of the house construction to his ongoing description of the hereditary process, relying on the former to both explain and support his explanation in the latter.

> Materials from various structures might be moved and shuffled together in the yard, yet pieces from the same source would frequently remain in juxtaposition. . . . So in the process of transmission by inheritance, elements derived from the same ancestor are apt to appear in large groups, just as if they had clung together in the pre-embryonic stage, as perhaps they did. They form what is well expressed by the word "traits," traits of feature and character—that is to say, continuous features and not isolated points. (8–9)

In these lines, Galton moves nimbly from narrative to explanation, drawing on the first to support his characterization of the second. In this case, the hypothetical tale of the brickyard, which depicts the author and the reader roaming about for building material to construct their house, helps explain to the reader both the random process of trait selection as well as the tendency of particulate traits to stick together as they pass from the brick yard of the gene pool into the biological structures of the next generation.

To emphasize the stochasticity of the transfer of material hinted at in the brick yard narrative, Galton goes on to explain:

> It would seem that while the embryo is developing itself the particles more or less qualified for each new post wait as it were in competition, to attain it. . . . Thus the step-by-step development of the embryo cannot fail to be influenced by an

> incalculable number of small and mostly unknown circumstances. (9)

In the final phrase, "an incalculable number of small and mostly unknown circumstances," Galton references Laplace's description of the types of causes that underlie probabilistic events whose outcomes distribute normally in a bell-shaped curve. This reference expresses Galton's opinion that the selection of traits to be expressed in the embryo are under the influence of the same kind of causes and can, therefore, also be considered probabilistic and open to assessment using the law of error.

He makes this argument explicitly at the close of the chapter where he writes:

> What has been said is enough to give a clue to the chief motive of this chapter. Its intention has been to show the large part that is always played by chance in the course of hereditary transmission, and to establish the importance of an intelligent use of the laws of chance and of statistical methods that are based upon them, in expressing the conditions under which heredity acts. (*Natural Inheritance* 17)

Whereas chapter two relies on a movement between hypothetical narrative and description to illuminate the process of heredity in the individual, chapter three offers the reader a broad, multi-generational vision of reproductive outcomes with the help of forensic narrative elements. This change of method supports Galton's claims for the stability and comparability of traits across members of a population. In the opening paragraph of the chapter, the focus shifts from the process of trait combination and selection in an individual to the distribution of traits within a population.

> The total heritage of each man must include a greater variety of material than was utilized in forming his personal structure. The existence in some latent form of an unused portion is proved by this power, already alluded to, of transmitting ancestral characters that he did not personally exhibit. Therefore, the organized structure of each individual should be viewed as the fulfillment of only one out of an indefinite number of mutually exclusive possibilities. His structure is the coherent

> and more or less stable development of what is no more than an imperfect sample of a large variety of elements (18).

In this opening explanation, a few elements are remarkable. First, Galton acknowledges that some traits are unexpressed, which reveals that he, like Mendel, had something like an idea of the distinction between the genotype, the total number of character traits in an organism, and the phenotype, the number of characters which are physically expressed by the organism. Second, he makes the case that the elements that are expressed are only some out of an indefinite number of mutually exclusive possibilities. This statement reveals that Galton was thinking along the lines of a gene pool of which individuals were only samples. This contrasts strongly with the traditional "essence" theories of species, such as the ones proposed by Kölreuter and Gärtner, in which variation was only deviation from an ideal type.

After making these general characterizations about the distribution of variation in populations, Galton offers specific arguments supporting his position. He begins by challenging the traditional theory of inherited similarity as the direct transfer of traits by describing in detail what he believes is an equally plausible gene pool scenario.

> It appears that there is no direct heredity relation between the personal parents and the personal child . . . but that the main line of hereditary connection unites the sets of elements out of which the personal parents had been evolved with the set out of which the personal child was evolved. (19)

In this statement Galton reveals that he is thinking of heredity in terms of populations rather than in terms of individuals because he sees parents and children as random expressions of characters out of a collection of traits. This characterization provides a rationale for accepting both the necessity of probability and statistics as tools for studying inheritance, and a rejection of the notion that characteristics change as they are passed down from one generation to the next. If heritability is seen as a fairly direct transfer of traits which are only slightly altered as a result of environmental or other factors, as Darwin suggests, then the phenotype of the parents and the individual offspring can suffice as the subjects for investigating inheritance. If, however, individual offspring are simply one of infinitely many combinations of a set of heritable traits, then an investigation of the features of single individuals becomes meaningless because it can tell us nothing about the larger pool of traits or the laws of selection of traits from that pool.

As a result, the study of inheritance by necessity becomes the study of large fraternities or substantial samples of related organisms from a single known heritage which are capable of revealing patterns in and ranges of character combinations of parental traits, a point also recognized by Mendel. This type of assessment, according to Galton, can only be accomplished through the use of statistical and probabilistic analysis.

> We are unable to see particles and watch their groupings, and we know nothing directly about them, but we may gain some idea of the various possible results [of the combination of character traits] by noting the differences between the brothers in any large fraternity . . . whose total heritages must have been much alike, but whose personal structures are often very dissimilar. That is why it is so important in hereditary inquiry to deal with fraternities rather than with individuals, and with large fraternities rather than small ones. (19–20)

Once Galton has described the process of inheritance in terms of random selection from a general pool of traits, he has to explain how people related to one another tend to pass down collections of features to their offspring. To account for this characteristic of the hereditary process, Galton turns to forensic narrative. In his explanation of how positions of stability can emerge from randomly selected traits, he tells the reader the story of the rocks of Kenilworth Castle. He begins: "The whimsical effects of chance in producing stable results are common enough. . . . Many years ago there was a fall of large stones from Kenilworth Castel" (21).

Offering his own recollection of the past event, Galton continues his story, testifying that "three of them [the stones], if I recollect rightly, or possibly four, fell into a very peculiar arrangement, and bridged the interval between the jambs of an old window" (21). At this point, he calls in nameless witnesses whose presences also support his testimony about this past event: "the oddity of the structure attracted continual attention, and its stability was much commented on. These hanging stones, as they were called, remained quite firm for many years; at length a storm shook them down" (*Natural Inheritance* 21–22).

In the tale of the hanging stones, Galton offers his readers a forensic narrative reminiscent of the observational commentary of seventeenth century science. He reports on an event which the reader may or may not have seen, and appeals to the experience of other witnesses as corroborating evidence of the existence of the phenomena. He then uses the instance as evidence to support his speculations on heredity: if some random incident of stability can occur in physical matter, it should also occur amongst the particles of heredity as they are selected over thousands of years.

In chapters two and three of the text, Galton blends forensic and hypothetical narrative with hypothesized descriptions of the heredity process to make his case that the warranting conditions necessary for his mathematical arguments exist. In his hypothetical descriptions, he explains how the formation of traits and their selection from the gene pool are probabilistic and uses the story of the brick yard to support and clarify his position. In addition, he uses the hypothetical narrative of the brick yard and the forensic narrative of the stones of Kenilworth Castle to explain how stability of trait compounds formed from a combination of trait particles can be maintained. The stability of features supports his position that hereditary traits can be passed down as collectives from one generation to the next and, like features of physical phenomena, can be considered fixed comparable instances, unaffected by causes that might seriously skew an assessment of their true value.

Chapters Four and Five: An Argument for Method

While Galton relies on narrative to establish the special warrants required for his mathematical argument in chapters two and three, in chapters four and five he turns his attention to arguing specifically for the applicability of the law of error to investigations of the hereditary process. These chapters involve particularly intricate arguments toward this end. On the one hand, Galton wants to make that case that the principles and formulae associated with the law of error are applicable to the study of heredity. On the other hand, he also wants to argue for a new way of interpreting the law of errors, which changes the focus of attention from the median value of a measurable trait to the deviation of trait values from the median. In order to persuade his readers to accept this new interpretation, Galton appeals to their Victorian sensibilities about the importance of knowing one's place or rank, and to rhetorical strategies of definition and visual argument.

In Chapter 4, "Schemes of Distribution and Frequency," Galton appeals to the values of his reader to make the case for the importance of statistical deviation. The law of error, as it was commonly used in physics, history, astronomy and sociology, was most often, if not exclusively, employed as a means of ascertaining the *mean* value of a feature in a population. Galton recognizes that this particular focus on the mean as the object of study is likely the one that his reader, if he has had any experience with the law of error at all, will recognize. In Galton's unique application of the law to the subject of inheritance, however, it is the *deviation* of a particular variation from the mean, not the mean itself, that is the object of interest.

Because the primary focus and goal of the traditional use of the law of error is to find the mean value in a distribution of values, Galton begins his attack on the orthodox application of the law by assaulting the notion that the mean value provides the researcher with the most significant information about a population. To challenge the importance of the mean, Galton offers an example from the financial world. He describes a case in which the average income of the English population is known, and explains that this information tells virtually nothing about the population itself. In contrast, a description of the deviation of a particular variation from the mean can reveal, "what proportion of our countrymen had just and only just enough means to ward off starvation, and what were the proportions of those who had incomes in each and every other degree" (*Natural Inheritance* 35–36). In other words, knowledge of the average value and the distribution of values allow individual values to be ranked or compared with all other values. Although the average value alone can also serve as a ranking index, it can only supply an idea of whether a value is above or below average, whereas knowledge of the distribution of values can reveal the extent to which the population does or does not share a similar value.

The concept of rank within the overall population is important for Galton, and most likely for his Victorian readers, who existed in a culture obsessed with the concepts social place and progress. Although Galton is a strict believer in the determinacy of natural endowment, rejecting much of the Victorian sentiment surrounding self-improvement, he makes a point at the end of his example to emphasize the importance of the concept of rank and its value in ascertaining social progress.[6]

> A knowledge of the distribution of any quality enables us to ascertain the rank that each man holds among his fellows, in

> respect to that quality. This is a valuable piece of knowledge in this struggling and competitive world, where success is to the foremost, and failure to the hindmost, irrespective of absolute efficiency. . . . When the distribution of any faculty has been ascertained, we can tell from the measurement, say of our child, how he ranks among other children in respect to that faculty, whether it be a physical gift, or one of health, or of intellect, or of morals. As the years go by, we may learn by the same means whether he is making his way towards the front, whether he just holds his place, or whether he is falling back towards the rear. Similarly as regards the position of our class, or our nation, among other classes and other nations. (37)

In these lines, Galton explains that knowledge of the distribution of trait values could provide the reader with a way of ascertaining how his children, with regards to these highly esteemed attributes, would measure up to other children as well as supply him with a means of tracking over time the improvement or decline in these attributes. In the competitive, late nineteenth century Darwinian environment, a method by which the rank of one's offspring could be ascertained and tracked would be of the greatest utility because, though abilities are determinate, it might provide knowledge about which child to risk time and wealth on, and which child such endowments might yield scant return. By characterizing the usefulness of knowing the distribution of values in terms of the ranking of offspring, Galton draws upon the values of the readers and their aspirations to make his case for the value of the law of error and to tie human biological and psychological phenomena to that law.

Once Galton has preemptively made the case for the importance of determining the range of values for a given trait in a particular population over the importance of knowing the mean value, he introduces the law of error in the fifth chapter of the text, titled, "Normal Variability." Here he presents arguments that the law of error can be applied to natural data, but that a distinction had to be made between the traditional applications of the law and the approach he is employing.

In the opening of the chapter, Galton introduces the concept of *deviation* and makes the case that when natural data are marshaled, their organization reveals regularity in terms of deviation from the median, which is similar to the regularity observed in the deviations from the median in the mathematically derived law of error. In the first section, "Normal Variability," Galton explains that the mathematically derived model for the probable distribution of errors reveals a "scheme" of deviations that is fixed and differs to some degree from the distribution of the values represented in his anthropometric data.[7] By mathematically "smoothing" the curve, however, Galton claims he is able to make the deviations of the anthropometric data fall nicely into line with the scheme derived from the normal law of error.

As evidence for this claim, Galton presents the values of deviation in the known tables of probability integrals (which describe the scheme of the law of error) side-by-side with the smoothed values for deviation in the anthropometric data for height, weight, arm span, etc. in his Table 3 (here Table 4) of the text. Using the table, he draws his reader's attention to the visual evidence of their quantitative similarity: "All the 18 schemes of deviation that can be derived from Table 2 have been treated on these [smoothed] principles, and the results given in Table 3. Their general accordance with one another, and still more with the mean of all of them, is obvious" (54).

With the table, Galton makes a powerful, rhetorical argument using visual comparison to support his conclusion that an analogy exists between the law of error and heredity. The comparison works in two ways to make his case. First, the mass of figures provides an ocular demonstration that the deviation from the sample mean replicates the expected deviation of outcomes described by the law of error. As explained earlier, the distribution of outcomes of a probabilistic event is described by the bell curve, which has the particular feature of the being symmetrical on either side and having a central, true value equal to zero (zero error). In this distribution, the values of error or deviation increase uniformly as they move away from the true value in the center but decrease in frequency.

A brief glance at the table is sufficient to see that the deviation values calculated by Galton using his anthropometric data distribute according to this scheme. In the very center, the reader is greeted with a long line of zeros representing the true values of all the physical features described. The zeros represent the true values because there is no error in their measurement. Then, moving out on either side, the values of the chart continuously rise in symmetrical perfection as the amount of the error rises. Without even carefully scrutinizing the data, it is possible for the reader familiar with the signature pattern described by the law of error to take in at a glance Galton's visual argument for the similarity between the distribution of heritable physical traits and the law of error.

Table 4. Values of mathematically smoothed anthropometric data compared with the mathematically derived values from the law of errors. Source: *Natural Inheritance*, pg. 201.

TABLE 3.

DEVIATIONS from **M** in each of the series in Table 2, after reduction to a Scale in which $Q' = 1$, where Q' is the *Mean* of the observed Deviations at the Grades 20°, 30°, 70°, and 80°.

Subject of measurement.	Values of Q'	Unit of measurement in Table 2.	Sex.	No. of persons	Deviations reckoned in units of Q'.										
					5°	10°	20°	30°	40°	50°	60°	70°	80°	90°	95°
Height, standing, without shoes . .	1·72	Inches	M.	811	2·73	1·98	1·22	0·81	0·35	0	0·35	·76	1·22	1·98	2·61
	1·62		F.	770	2·71	2·10	1·23	·74	·37	0	·37	·80	1·23	1·91	2·46
Height, sitting, from seat of chair	0·95	Inches	M.	1013	2·52	1·89	1·15	·73	·63	0	·31	·73	1·15	1·79	2·31
	0·82		F.	775	2·55	1·95	1·22	·73	·36	0	·36	·85	1·22	2·07	2·55
Span of arms . . .	2·07	Inches	M.	811	2·36	1·83	1·30	·82	·43	0	·33	·72	1·16	1·79	2 36
	1·87		F.	770	2·35	1·87	1·23	·69	·32	0	·37	·80	1·28	1·98	2·67
Weight in ordinary indoor clothes . .	10·00	Pounds	M.	520	2·20	1·80	1·20	·80	·40	0	·40	·70	1·30	2·20	2·90
	11·00		F.	276	1·80	1·60	1·10	·70	·40	0	·60	·90	1·30	1·80	2·40
Breathing capacity .	24·50	Cubic	M.	212	2·32	1·68	1·28	·80	·32	0	·28	.68	1·16	2·32	2·84
	19·00	Inches	F.	277	2·39	1·87	1·20	·73	·36	0	·31	·67	1·35	2·03	2·49
Strength of pull as archer with bow .	7·50	Pounds	M.	519	2·39	1·86	1·33	·80	·40	0	·40	·80	1·06	1·99	2·92
	5·22		F.	276	1·92	1·06	·80	·53	·27	0	·27	53	·93	1·46	1·86
Strength of squeeze with strongest hand	7·75	Pounds	M.	519	2·32	1·81	1·16	·77	·39	0	·39	·77	1·29	1·93	2·45
	7·50		F.	276	2 12	1·73	1·20	·66	·40	0	·40	·80	1·33	1·99	2·66
Swiftness of blow	2·37	Ft. per second	M.	516	2·06	1·68	1·22	·80	·34	0	·42	·80	1·18	1·77	2·31
	1·55		F.	271	2·71	2·13	1·35	·84	·38	0	·38	·71	1·10	1·87	2·26
Sight, keenness of — by distance of reading diamond test-type	4·00	Inches	M.	398	3·00	2·00	1·25	·75	·50	0	·25	·75	1·25	1·75	2·25
	5·22		F.	433	2·66	2·28	1·52	·95	·38	0	·38	·57	·95	1·33	1·52
SUMS . .					43·11	33·12	21·96	13·65	7·00	0	6·57	13·34	21·46	33·96	43·82
MEANS . .					2·40	1·84	1·22	0·76	0·39	0	0·37	0·74	1·19	1·89	2·43
MEANS multiplied by 1·015, to change unit to $Q = 1$					2·44	1·87	1·24	0·77	0·40	0	0·38	0·75	1·21	1·92	2·47
Normal Values, when $Q = 1$					2·44	1·90	1·25	0·78	0·38	0	0·38	0·78	1·25	1·90	2·44

Though Galton uses the arranged data on the chart to make the case for a pattern of similarity between the distribution of values of physiological phenomena and the law of error, he does not stop at arguing for an approximate relationship between the two. Instead, he pushes his case further using visual comparison between the mean values of the physical features in each percentile group and the value calculated for the distribution of traits in that group, using the law of error to show that there is almost an exact correlation between the observed values in his anthropometric data and the values calculated by the law of error.

Galton makes the visual comparison between the two for the reader in the last two rows of the table, labeled "Means" and "Normal Values." In the second to last row, "Means," Galton mathematically transforms the means of the anthropometric data into a unit comparable to Q, the standard value of probable error according to the law of error, by multiplying the raw data, Q', by the value 1.015, the average in the mean deviations between the 20° and 30° grades and the 70° and 80° grades. This, according to Galton, transforms the actual values of the raw data Q', which are in larger units, to the smaller units of the standard value of probable error, Q (54). Galton then compares the transformed values, revealing that the distribution of the actual data follows very closely with the ideal distribution of probable error.

The close similarities between the distribution of and the values for deviation in the natural data, and the distribution and values calculated from probability, provide sufficient evidence that the mathematical formulae and principles that have been developed for describing the law of error can also be applied to describing the distribution of natural data. By choosing an appropriate scheme and arranging the data of calculated deviations from the mean on the table according to that scheme, Galton relies on visual tactics to persuade readers to accept his proposed analogy between hereditary outcomes and the distribution of values according to the law of error.

Although Galton's visual argument makes the case that the law of error is mathematically suitable for describing *deviations*, or errors, from a true value in anthropometric data, he has not made the case that that they are suitable for describing *variability*, which is the focus of his hereditary investigation. As a result, he still needs to persuade his reader that these deviations can be interpreted in terms of natural variation. To accomplish this, Galton makes an argument using the rhetorical figure *synonymia* (also known in Latin as *interpretatio*), with which he attempts to establish a general similarity between the concepts of error and variation. According to the standard application of the law of error, the values of the observational results that fall above or below the mean "true value" represent a distribution of erroneous observations. By calculating the deviation of possible errors from the median, it is possible to estimate how far away in contrast with other measurements from the "true value" a particular measurement lies and, therefore, calculate the probability of its erroneousness. As a result, deviations describe, in the nomenclature of the law of errors, "probable errors."

To equate "error" with "variation" Galton uses the rhetorical figure *synonymia*, a chaining together of synonyms meant to amplify or explain a given subject:

> But errors, differences, deviations, divergencies, dispersions, and individual variations, all spring from the same kind of causes. Objects that bear the same name, or can be described by the same phrase, are thereby acknowledged to have common points of resemblance, and to rank as members of the same species, class, or whatever else we may please to call the group. (55)

Here Galton strings together synonyms to move from "errors," the commonly used term in the epistemic culture of the physical sciences, to "variations," the term typically used to talk about the biological phenomena of evolution and heredity. The connection of these terms by intermediary synonyms suggests to the reader that errors and variations can be grouped with similar terms and, therefore, must themselves be synonymous.

In addition to using synonymia to connect "errors" and "variations," Galton employs the figure to make the case that the two are effects of the same cause: "the combined influence of a multitude of accidents" (55). From this conclusion, he argues that the mathematics which applies to the calculation of errors, therefore, should also reasonably apply to the calculation of variability.

> All persons conversant with statistics are aware that this supposition [that errors, differences, etc. are all the result of a multitude of small accidents] brings variability within the grasp of the laws of chance, with the result that the relative

> frequency of deviations of different amounts admits of being calculated, when those amounts are measured in terms of any self-contained unit of variability. (55)

Although Galton makes the case for identity between the terms "error" and "variation" for the purpose of suggesting to the reader that the law of error can be used to describe variability, he is quick to differentiate his use of the law of error to study variability from the traditional applications associated with it. To make this distinction, Galton relies on arguments from definition. He begins with the conclusion that because the science of heredity has different ends than those of mathematics, the specific manner in which terms are employed by the one is not particularly useful to the other.

> It has already been said that mathematicians labored at the law of error for one set of purposes, and we are entering into the fruits of their labors for another. Hence there is no ground for surprise that their nomenclature is often cumbrous and out of place when applied to problems of heredity. This is especially the case with regard to their term of "probable error." (57)

Although initially Galton has no quibble with the law of error's concept of "probable error" (Q) and describes it as co-referential with variability, at this point in the text he characterizes the term as misleading and in need of proper clarification. He argues that in common parlance, most people would take the phrase "probable error" to mean the most probable error which, mathematically, would be the median error. The median error, however, is not an error at all, but instead the "true value," or the case of zero probable error. Because the term is so misleading, he claims that his recasting of the statistical phenomena of "probable error" as "probable deviation" is not only justified, but necessary. In this way, Galton is able to separate his terminology from the terminology commonly employed in the law of errors while at the same time maintaining the position that, conceptually, the law of errors is compatible with the study of variability.

After five chapters of carefully laying the foundation for the law of error using rhetorical strategies, Galton proceeds in chapters six through eight of *Natural Inheritance* to offer the empirical evidence and mathematical proof required to make his case. In combination, chapters one through eight represent a deliberate and well-orchestrated argument for a mathematical approach to the study of heredity. Within these chapters, Galton builds his argument incrementally using a variety of strategies, including rhetorical, empirical, and rational. In chapters two through five, he employs rhetorical strategies such as narrative, visual argument, synonymia, definition, and appeals to Victorian values to persuade his readers to accept his warrants, his characterization of the law of error, and his application of the law to the study of heredity.

In chapter seven he argues mathematically for the relationship between heredity and the law of error, and from that relationship, to establish the law of regression. Finally, in chapter eight, he uses the law of regression deductively to predict the distribution of eye-color in family data, further verifying the connection between the natural process of heredity and the mathematical law of error.

Although the evidence and methods used in the later chapters had a substantial impact on the audience's acceptance of Galton's claims, without the initial rhetorical arguments they might have been received with a greater degree of skepticism, as the data and the mathematical method in *Hereditary Genius* had been. As we will see from evidence of the book's reception, these strategies, in addition to Galton's ethos, had a substantial influence over the success of the text.

The Reception of *Natural Inheritance*

A close textual analysis of *Natural Inheritance* suggests that Galton relies on rhetorical tactics to argue his mathematical method into place; however, it does not supply proof that these strategies elicited the desired response from his readers. In order to make the case (1) that at least some members of his audience were persuaded of the legitimacy of his arguments, and (2) that their positive opinions were encouraged by the strategic rhetorical choices made by Galton in *Natural Inheritance*, this section examines reviews of the book and the changes in research trends after the book's publication.

A search of newspapers, magazines, and journals immediately following the book's publication yielded a total of eight reviews of Galton's work. Half of the reviews were published in popular newspapers and magazines, including the *Times*, the *New York Times*, the *Spectator*, and *Nation*. The other half of the reviews were published in journals, including inter-specialist scientific journals, *Nature* and *Science*, and intra-specialist journals, *Mind* (a journal of philosophy) and the *Publications of the American Statistical Association.*

Despite the heterogeneity of sources, there is a consensus among the reviewers that *Natural Inheritance* was an important and substantive text. Out of the eight reviews, seven are positive. A comment by an anonymous reviewer in the *New York Times* on April 14, 1889, offers a glimpse of the type of positive comments made about the book: "For perfecting philosophical inquiry, for prudence and good judgment, Mr. Galton is to be considered as presenting the highest examples of modern research" ("From Father to Son" 19).

In addition to being generally complementary, at least five of the seven positive reviews stated that Galton's work was ground-breaking, a new science, or a novel method of research.[8] Testifying to the novelty of Galton's work, an anonymous reviewer in *Science* writes, "A prominent feature in the present work is the application of the 'probability curve' to the facts of physical variation,—an attempt to apply mathematical conception in the field of biology, and to found a science of biological statistics" ("Natural Inheritance," 266).

The only negative review of the text was published in *Nature* by psychologist/philosopher Hiram M. Stanley of Lake Forest University. Stanley attacks the Galton's probabilistic conclusions on the grounds that they assume that the cause of variation comes only from heredity, ignoring any possible influence of external environmental factors.

> As to the way in which an abstract calculation of the laws of chance confirms Mr. Galton's statistics, it is enough to observe that no evidence is adduced why the results attained should not stand for the multiple "accidents" of environment . . . rather than the "accidents" of heredity alone. Mr. Galton fails to prove that his ratios are not the mathematical expression for the operation of heredity plus other agencies, rather than the formula for heredity simple and unadulterated. (643)

Stanley's complaint echoes protests of Galton's earlier efforts in *Hereditary Genius* to make claims for the normal distribution of intelligence amongst the scores of Cambridge mathematics students.

The positive response by the majority of reviewers to *Natural Inheritance* and their characterization of it as trailblazing text supports the supposition that, generally, Galton's arguments were compelling for his readers. Evidence of the general success of the text, however, provides no proof that any of the particular strategies—narrative, analogy, appeal to the values of the audience, empirical verification, mathematical verification, and prediction—or his ethos were factors in audience persuasion. Further scrutiny of the reviews suggests that they were, though every strategy was perhaps not effective for every reviewer.

In the very first section of this chapter, evidence of Galton's stature within the scientific community was presented, and the claim was advanced that Galton's ethos contributed to the success of his text with his audience. An examination of the reviews of *Natural Inheritance* supports this conclusion. In five of the seven positive reviews, the reviewers made mention of Galton's status either specifically, as a scientist with a long history of research on the subject of heredity, or generally, as a scientist with a sterling reputation (*Spectator*, *Times*, *Science*, *New York Times* and Venn). In the *Times*, for example, an anonymous reviewer reminds readers of the Galton's past work in heredity as a means of supporting his own conclusions about the general value of Galton's work: "Mr. Galton began is studies on heredity many years ago; and he may still boast that he conducts his inquiries by more refined and searching processes than those hitherto employed" ("Mr. Galton" 3).

A more spectacular example of the influence of Galton's situated ethos on the reception of the text can be found in a review in *Science*, in which an American reviewer offers an opinion on Galton's general reputation as a gentleman of science:

> Francis Galton hardly needs an introduction to American readers. His research into the heredity of genius, his study of predominant traits of English Scientists, his invention of composite photographs, together with a large number of interesting and original memoirs, have made his name and work known wherever new applications of scientific methods are appreciated. (266)

In addition to his scientific reputation, Galton's rhetorical tactics in *Natural Inheritance* played a role in persuading audiences to accept the arguments in the book. Interestingly, though perhaps not surprisingly, Galton's strategy of using narrative in chapters two and three were most often commented upon by reviewers in the popular press. These reviewers—perhaps because of their limited expertise, or perhaps because of their perception that their readers would have limited expertise—are drawn to Galton's non-quantitative arguments and use them to rationalize or explain the mathematical arguments in the text. The anonymous reviewer in *Nation*, for example, provides a rationale for Galton's mathematical "law of regression" by alluding to the first section of the text where Galton uses the narrative of the stones of Kenilworth Castel to make a case for the possibility of organic stability:

> It might be well to say a word here as to the reason why this phenomenon of regression might be expected *a priori* to take place. Two such reasons may be given. . . . The first is that unless such a tendency to regression existed deviation from the original type of the race would rapidly become more and more exaggerated, a state of things which is generally contradictory to the doctrine of organic stability—a doctrine which is somewhat vaguely but very suggestively discussed in the introductory part of Mr. Galton's book. (196)

In these lines, the reviewer in *Nation* makes a point to identify the introductory section of *Natural Inheritance* as the source for Galton's biological theory of organic stability. He then explains that this theory is the foundation on which Galton rests his law of regression, which is presented mathematically in chapter seven of the text.

Although the reviewer in *Nation* accepts Galton's biological narrative as a compelling rationale for his deductive argument, another critic in the *Times* conveys his skepticism about whether the narrative is sufficiently robust to be persuasive. He has some especially harsh words for the analogies employed in chapter three that liken the accumulation of traits in individuals over generations to practices in which parts of old buildings are used to make new ones. After describing Galton's conclusions in chapter three, the reviewer remarks:

> What is Mr. Galton's way of looking at the problem [of heredity]? An illustration will be the best answer. You look at an old wall in Rome or Ravenna. . . . The corner stone is a lintel borrowed from a temple; here is a mass of masonry riven in a block from a prostrate wall. . . . This composite structure, new and yet so old, is faintly symbolic of what we all are. . . . Such is the obscurity of the subject that Mr. Galton, at times the most scientific of inquirers, slides often into a half poetic, half mystical vein: he flies to parables, hypotheses, and ingenious

> are his efforts to cope with the subtlety of the facts. . . . There is occasionally the same tendency [as in the writing of Lucretius the Roman poet and philosopher] to suffer conclusions to outstrip facts and to mistake plausible simile for syllogism. ("Mr. Galton" 3)

In addition to the strategies in chapters two and three, Galton's rhetorical tactic in chapter four, of using Victorian values of rank to support the importance of examining the variation of a trait over its mean gets some positive attention from reviewers. In particular, it is recognized and accepted by the American philosopher John Dewey in his review of the book in the intra-specialist journal, *Publications of the American Statistical Association*. In his discussion of *Natural Inheritance*, Dewey seems to chide his statistical readers for not considering the importance of the distribution of errors using Galton's arguments:

> The keynote in the statistical side of the work is contained in Galton's statement that statisticians are apt to be content with averages, while an average is only an isolated fact. What is wanted is a method of *calculating distribution*. . . . For example, compared with the knowledge of the average income of an English family, a knowledge of how the total income of England was distributed would be much more important. (331)

In addition to be being persuaded by Galton's rhetorical arguments, Dewey also accepts Galton's claim of the analogy between probability and heredity on the basis of the empirical evidence he had collected. In his discussion of chapter five—where Galton compared the distribution of a variety of features such as strength, weight, arm span, etc. to the distributions calculated from the law of error—Dewey writes, "Galton found a remarkable parallelism between the results obtained by observation and those theoretically deduced from the mathematical calculations. . . . The results differ but little from those theoretically obtained from the law of frequency of error. It is evident that Galton might express his scheme in the well-known curve of error" (333).

Like Dewey, writers in the popular media recognized and expressed satisfaction with Galton's disciplined use of conventional argument strategies such as mathematical proof and prediction for making the case for the correlation between heredity and the law of error. The anonymous reviewer in *The Spectator*, for example, recognizes and is persuaded by Galton's geometrical proof of regression in chapter seven.

He writes: "In the chapter on stature, Galton's method is proved by its yielding the same results (barring small errors) as a purely mathematical treatment of the tables" ("The Scientific Treatment of Statistics" 84).

A brief examination of the reviews of *Natural Inheritance* suggests that Galton was generally successful in persuading his audience to accept the analogical connection between the probabilistic law of error and heredity, and to agree with his deductions of phenomena, such as regression, from that analogy. It also reveals that the specific rhetorical strategies identified in the text through close reading were recognized and had a positive impact on Galton's reviewers. Though not all of the strategies were equally productive of consent, the fact that they all had some degree of persuasive success suggests that we might fairly claim that these specific features had some broad positive influence on Galton's readership.

This assumption is corroborated by evidence that the book itself inspired some members of the scientific establishment studying biological phenomena to adopt a Galtonian approach to the study of variation and heredity. Perhaps the most important figures to be inspired by the book were the biological researcher Frank Raphael Weldon and the mathematician Karl Pearson. In collaboration, Weldon and Pearson established the biometrical school of biology, which endeavored to explore the phenomena of variation, evolution, and heredity from a mathematical standpoint. With full financial and restrained intellectual support from Galton, they founded the journal *Biometrika*, which was totally dedicated to pursuing biological questions from a mathematical perspective. Though Weldon died unexpected in 1906, Pearson continued to advance Galton's legacy using a financial bequest from Galton after his death in 1911 to found the Eugenics Laboratory at University College London.

In the first decade of the twentieth century, the clashes between the biometricians, headed by Pearson and Weldon, and the Mendelians, led by biologist William Bateson, who was also influenced by Galton's work, helped shape the path of biology. Though the biometricians lost the theoretical battle to Bateson and the Mendelians, their statistical practices remained robust methods for research and served as the catalyst for the emergence of mathematical population genetics in the 1920s and 30s.

Looking back over the decade preceding the turn of the century, Charles Davenport—in a report from the Committee of the American Association for the Advancement of Science on the Quantitative Study of Variation—writes in *Science* (1900) that Francis Galton and his work in *Natural Inheritance* represented the single most important influence in the development of the quantitative study of variation: "The beginning of the new decade [the 1890s] saw the beginning of a wider interest in the quantitative study of variation, and the source of this wider interest can be traced directly to one man—*Francis Galton*" (866).

Conclusion

Unlike Darwin, who employed the well-accepted common warrants of quantification and arithmetical comparison to invent and support his arguments in *The Origin of the Species*, Galton's mathematical arguments in *Natural Inheritance* employed special warrants about nature that required further arguments to establish. A close examination of chapters two through five of *Natural Inheritance* reveals that rhetorical strategies play a significant role in establishing the viability of the mathematical arguments in the text. In chapters two and three of the book, Galton relies on narrative and analogy to make the case for the validity of the warrants supporting his mathematical methods by framing the heredity process as probabilistic and hereditary traits as stable and comparative across members of a population. In chapters four and five, he appeals to the values of the audience to convince them that the variation of a population is more significant than its mean, and uses synonymia and definitional arguments to make the case that the law of error was applicable to the study of heredity yet different from it.

The effectiveness of these rhetorical strategies, in conjunction with the empirical and logical arguments in the text, in molding the opinion of Galton's readers is evidenced by the reviews of the book. These suggest that each of Galton's strategies played some role in persuading readers to accept his argument. In combination, evidence from historical context, close textual analysis, and audience responses to the text challenges the perception that mathematical concepts and formulae are accepted without question as self-evidentiary warrants for arguments about nature, and suggests that rhetorical arguments from ethos, narratives, and values can play a central role in the acceptance of mathematical arguments into the epistemic commitments of scientific communities.

6 Behind the Curve: Karl Pearson and the Push for Theoretical Mathematical Biology

> *The committee you have got together is entirely unsuited to direct such experiments. It is far too large, far too much of the old biological type and is far too unconscious of the fact that the answers of the problems required are in the first place statistics and in the next place statistics and only in the third place biology.*
>
> —Karl Pearson to Francis Galton, 1897

With the help of the rhetorical strategies in *Natural Inheritance*, Galton successfully persuaded at least some of his audience to accept his analogy between the hereditary outcomes and the mathematical law of error. The acceptance of this analogy led to the broader development of a statistical and probabilistic approach to heredity, which was adopted by a small but important group of young natural researchers in the last decade of the nineteenth century.

During this period, three of Galton's most notable disciples—Karl Pearson (1857–1936), William Bateson (1861–1926), and Frank Weldon (1860–1906)—increased the number of biological phenomena investigated using Galton's techniques and developed novel mathematical applications for biological research. Though they all drew inspiration from Galton's work, they eventually became bitter intellectual rivals. Historical accounts of the divisions between the Mendelians, led by Bateson, and the biometricians, fronted by Pearson and Weldon, suggest that the main sources of conflict between them were differences in their models of heredity (Provine 89). This chapter, however, broadens the investigations of these differences by focusing on the conflict over the role of mathematics as a tool for investigating heredity.

Whereas Galton had to persuade biological researchers to accept the validity of his mathematical approach to heredity, participants in the biometric controversy took for granted the legitimacy of Galton's method. What became an issue instead was whether the investigation of heredity should be guided exclusively by mathematical methods or if a mix of approaches was a better way to pursue knowledge about variation, evolution, and heredity. As a consequence, the debate about mathematics in science moved generally from the stases of existence and definition (Could mathematics be used to describe heredity, and are variation and error synonymous?) to the stasis of value (Is a mathematical approach to heredity the best approach for creating true or exact knowledge about heredity?) By illustrating that mathematical arguments can exist at the stasis of value, this chapter illuminates another rhetorical dimension of mathematical argument.

Method

As in previous chapters, this chapter employs a number of methods to establish the relationships between science, mathematics, and rhetoric. In the first section, rhetorical theory and historical methods provide a framework for understanding *value* as a concept and for identifying the values that predominated in the study of variation, evolution, and heredity in the last decade of the nineteenth, and the first decade of the twentieth centuries. To define *value*, the investigation turns to Perelman and Olbrechts-Tyteca's notions of the *loci of the real* and the *loci of the preferable* in the *New Rhetoric.* For an understanding of the values for late nineteenth and early twentieth-century arguers, it examines both the general trends in the scientific debate over heredity during this period as well as the particular philosophical commitments of the major participants in the Mendelian-biometric debate: Karl Pearson, Frank Weldon, and William Bateson.

In the second section, historical background and textual analyses of arguments from the Mendelian-biometric debate reveal the manner in which and reasons for which mathematics was elevated by the biometricians as an ultimate value in the study of variation, evolution, and heredity. These methods also illuminate how this particular value position was challenged by the Mendelians. To illustrate the centrality of mathematical methods as a value in his scientific program of research, the second section offers a close textual examination of Karl Pearson's editorial comments in the early issues of *Biometrika* and his arguments in "On the Principle of Homotyposis and Its Relation to Heredity." To understand the challenge to the biometrical position, the second section concludes with a discussion of William Bateson's response to "On the Principle of Homotyposis" in the *Proceedings of the Royal Society.*

Arguments from Values

To make the case that mathematics assumed the status of value in the Mendelian-biometric controversy, it is necessary to characterize the qualities of a value argument. The term "value," when commonly applied to argument, is often synonymous with "virtue" in the sense that a value is a quality that is held in esteem by a particular audience. For example, we might say that a political arguer is appealing to the values of the audience when he or she argues against big government (the virtue being freedom of choice) or for more cooperation with foreign powers (the virtue being communalism). However, the characterization of an argument as a *value* argument can also be made on the grounds that the position taken is recognized to enjoy only limited agreement amongst specific groups of constituents. Arguing, for example, against big government resonates with a very clear constituency, but it could by no means be accepted as a universal position and would be rejected by a variety of groups depending on the context and subjects to which its applied. While both of these perspectives on value are important to this investigation, the second use of the term "virtue"—a position with a limited agreement—is particularly germane to the discussion of the period before the Mendelian-biometric debate. During this period, the central ideas about the processes of variation, evolution, and heredity were generally divided into a series of binaries, neither of whose sides, at least initially, was generally considered more virtuous than the other.

In *The New Rhetoric*, Perelman and Olbrechts-Tyteca provide a theoretical foundation for understanding value argument as an argument with a limited scope of agreement. In their discussion, they separate argument into two main types: the *loci of the preferable*, which includes value arguments, and the *loci of the real*, which contains appeals made from facts, truths, or presumptions. The classification of an argument as an appeal to the *loci of the real* or the *loci of the preferable* depends on the arguer's belief about whether or not an appeal would be universally compelling for all of the members of the audience he/she intends to address.

To explain the quality of general appeal, Perelman and Olbrechts-Tyteca draw on the concepts of *agreement* and the *universal audience.* According to the authors, *agreement* is the starting point of argumentation. It is the perceived common ground of premises and rational operations connecting those premises that an arguer believes his/her audience already accepts. These perceived-to-be-accepted data and warrants provide the starting point for making decisions about what can be argued and how it can be argued (65).

The *universal audience* is the hypothetical audience the arguer believes would agree with the data and warrants he/she has chosen, and who would, therefore, likely be compelled to accept the conclusions arrived at using those data and warrants. By making the universal audience a hypothetical construct of the speaker rather than an actual set of persons, Perelman and Olbrechts-Tyteca address the problem of agency in the relationship between the arguer and the audience. In the models of demonstrative, self-evidential reasoning, which the authors of the *New Rhetoric* reject, the audience is without agency in the sense that if argument follows the strict parameters of proper demonstration, they are forced to accept it as legitimate or be considered irrational. The hypothetical nature of the universal audience, however, returns agency to the audience because it allows the arguer to fail at persuasion and makes the audience the final arbiter of whether an argument is or is not compelling. At the same time, however, Perelman and Olbrechts-Tyteca are careful not to eliminate altogether the notion that there can be reliable criteria for rationality. They accomplish this by keeping open the possibility that arguers can develop a familiarity with the parameters of argument that their real audiences find compelling, and thereby develop strategies for argument that would be effective. Because the universal audience can involve both the imagination of the arguer and the beliefs and practices of an actual audience being addressed, success or failure of argumentation revolves around the successful coordination between the agreements the arguer believes the audience shares, and the agreements actually maintained by the audience.

The difference between the arguments from the *loci of the real* and the *loci of the preferable* is in the extent of the commitment the arguer assumes a particular set of agreements will have for the imagined universal audience. In the case of the *loci of the real,* Perelman and Olbrechts-Tyteca explain that the arguer assumes that an agreement is unanimously accepted by the universal audience: "everything in argumentation that is deemed to relate to the real is characterized by a claim to validity vis-à-vis the universal audience" (66). Within the *loci of the real* reside *facts, truths,* and *presumptions,* all of which are thought to be universally accepted, though these agreements vary in the degree of certitude designated to them. *Facts* are data about objects, either observed or supposed, and *truths* are complex systems, such as scientific theories, which relate facts (67–69). The hallmarks of facts and truths are that they are certified agreements, assumed to be true. Therefore, the arguer assumes that no defense or corroborating evidence is required to support them. *Presumptions,* on the other hand, are unlike facts and truths because their certitude can be reinforced by evidence (70–71).

In contradistinction to the *loci of the real,* the *loci of the preferable* are agreements that the arguer does not expect to be binding on all members of the imagined universal audience:

> Besides facts, truths and presumptions, characterized by the agreement of the universal audience, our classification scheme must also find a place for objects of agreement in regard to which only the adherence of *particular* groups is claimed. These objects are the values, hierarchies, and *loci* of the preferable. (74, emphasis added)

Perelman and Olbrechts-Tyteca's characterization of value arguments as positions adhered to by particular groups is a valuable perspective from which to assess scientific argument in periods when there is no prevailing scientific paradigm. In *The Structure of Scientific Revolutions,* Kuhn provides a general scheme for characterizing the stability of science based on the breadth of agreements that members of a scientific, epistemic community identify as either facts or truths. He divides sciences into pre-paradigmatic, paradigmatic, and revolutionary stages. In the paradigmatic stage of science, there is a substantial body of agreed upon of facts and truths that provide reliable guidance for the prosecution of research. In the states of pre-paradigmatic and revolutionary science, however, no broad foundation of agreed upon facts and truths exist about particular natural phenomena to encourage and support the creation of a broad coalition of researchers working within the same parameters.

> Being able to take no common body of belief for granted, each writer on physical optics felt forced to build his field anew from its foundations. In doing so, his choice of supporting observation and experiment was relatively free, for there was no standard set of methods or of phenomena that every optical writer felt forced to employ. (13)

Without a paradigm, assessing value as virtue is a complex affair because there is often insufficient agreement that a particular theoretical perspective or method has a sufficient critical mass of constituents to be identified as virtue within the bricolage of theories, methods, and evidence that characterize pre-paradigmatic science. By characterizing value as division, however, pre-paradigmatic science becomes more amenable to analysis. The lines between different perspectives are easily identified and there is no requirement for the analyst to judge the point at which an idea has gained a sufficient following to be considered a virtue in contradistinction to some other virtue. Further, by characterizing value as division rather than virtue, premises or warrants that are considered equally credible—as is often the case in pre-paradigmatic science—are open for assessment where they might otherwise be overlooked because they do not have the characteristics of virtues. Because of these characteristics, the notion of *value as division* will be used in the initial characterization of the theories about and methods for investigating heredity in the nineteenth century.

Presumptions and Preferences: A Prelude to the Biometric and Mendelian Controversy

In the early part of last decade of the nineteenth century, the study of heredity was still a pre-paradigmatic science. This status is evidenced by the wide variation of theories and methods that existed contemporaneously with one another. For example, contradictory characterizations of the hereditary process, such as the belief that organisms could have some features that varied continuously, and other features that were discontinuous, were generally accepted to both be reasonable. Further, there was a more eclectic view towards methodologies for investigating hereditary phenomena, permitting Galton's statistical methods for investigating heredity to coexist with traditional, experimental hybridization methods. This section examines the personal commitments of Karl Pearson, Frank Weldon, and William Bateson, whose values and arguments shaped the debate between the biometricians and the Mendelians. This investigation shows that, until the latter half of the last decade of the nineteenth century, there was a reasonable degree of tolerance for different methods for investigating heredity amongst them. However, it illuminates the value divisions, which by the turn of the century would be advanced as value hierarchies to defend and define particular positions.

Karl Pearson

While Francis Galton is considered the conceptual father of biometry, Karl Pearson might be best described as its first radical disciple. With his zealous (and some would say single-minded) efforts and considerable mathematical talent he became a major force, along with Frank Weldon, in developing and promoting the use of mathematical models to describe the action of evolution, variation, and heredity in populations of organisms. A brief foray into Pearson's background suggests that this passion to create a biometrical paradigm was fueled by his dedication to mathematics, and perhaps more importantly, by his dedication to a unique philosophy of probabilistic positivism.

From very early in his academic career Pearson was directed by his father, William Pearson, to devote himself to the study of mathematics. In 1875 his father hired a famous tutor, E.J. Routh, to prepare Pearson for competition in the prestigious Cambridge mathematical Tripos, a nine-day mathematical examination whose top scorer garnered perhaps the highest academic prestige that could be bestowed on an undergraduate at Cambridge (Porter 22, 40). Despite dutifully applying himself to his mathematical studies with Routh, however, Pearson's efforts didn't pay off as he had hoped. After nine grueling days of examinations in the bitter cold of January, 1879, he finished third in the mathematical Tripos.

Despite what he perceived as a personal failure to establish himself as first-class mathematician, Pearson continued to pursue academic mathematical research. His main source of interest was in the mathematics of the ether, a hypothesized semi-elastic substance within which all molecules were supposed by some theorists to be suspended, and which reacted to their physical movements by telegraphing mechanical waves across space. The philosophical commitments that informed Pearson's model of ether, as well as the manner in which he carried out the development of his physical theory, offer insight into his epistemological commitments to developing and promoting a purely mathematical approach to the study of heredity.

In his efforts to create a physical model that explained the phenomena of gravitation, electricity, and light, Pearson sought to correct what he believed were the metaphysical shortcomings of mainstream physics. Unlike most prominent physical researchers at the end of the nineteenth century, Pearson was a positivist. He believed that originating causes of physical effects could not be known, and was committed to the idea that sense perception was the ground of all knowledge. These positions conflicted with prevailing physical theories that relied on unobservable properties of matter such as force and energy to explain its behavior.

In his theory of ether, for example, Pearson attempted to replace what he believed were the loose ideas of physics, such as force and energy, with more rigorous categories of mechanical motion, such as rate of change in location over time and vector. What he proposed was a study of the ether akin to hydrodynamics, which sought to explain all of the qualities of the physical universe as consequences of the mechanical motion of specific portions of the ether—he called these "atoms"—that were indestructible and contained both "internal" and "external" energy (Pearson, "Atomic Hypothesis" 71).

To construct his new theory, Pearson began his work not through observation of or experimentation with phenomena, but where he was methodologically most comfortable, in mathematics. Like most physicists of the time, Pearson pursued the development of his physical model using complex algebraic functions and the principles of geometry to describe the pulsation of the ether. With the help of these tools, he described gravity, molecular forces, and even the spectra for polyatomic molecules.

Pearson's highly theoretical and complexly mathematical approach to physical phenomena did not get much support from the physics community. Not only did it run philosophically contrary to some of the most important work in theoretical physics at the time by Maxwell, Thomson, and Hertz, but it also lacked empirical evidence. Pearson's problem of evidence is the subject of a letter written by George Gabriel Stokes, Lucasian Professor of Mathematics at Cambridge, in which he warned Pearson "that a correspondence of his deductions to 'results of actual observation in spectral analysis' would not prove any genuine resemblance between the systems he imagined and those 'actually existing'" (qtd. in Porter 187). Despite Stokes's well-intended critique of his work, Pearson continued to publish his mathematical theoretical papers on the subject until they ceased to be accepted by mainstream publications.

Though his physical theories never got off the ground, Pearson was persistent in his pursuit of positivistic science. In the early 1890s, he was laboring over the manuscript of *The Grammar of Science* (1892), a text whose purposes were to critique the fundamental goals of science and to reconstruct the rational principles of proper scientific methodology. Thought the text is predictably positivistic, with sections such as "Sensations as the Ultimate Source of the Materials of Knowledge" and "The Futility of Things-in-Themselves," they also offer important information about a unique facet of Pearson's positivist philosophy, which has important ramifications for understanding the direction of his later work in biometrics.

In "Cause and Effect as the Routine of Experience," Pearson begins with the positivist platform on the impossibility of knowing the final causes of, and on the necessity for, science to be based only on phenomena that can be observed by the senses. He explains that a cause or causes could only be understood as an instance in a discrete series of experienced events occurring one before another:

> Scientifically, cause, as originating or enforcing a particular sequence of perceptions, is meaningless—we have no experience of anything that originates or enforces something else. Cause, however, used to make a stage in a routine, is a clear and valuable conception, which throws the idea of cause entirely into the field of sense-impressions, into the sphere where we can reason and reach knowledge. (*Grammar*, 1892, 153)

What follows represents a novel aspect of his positivistic platform. Not only does he deny the possibility of knowing causes that are beyond our senses, but he also rejects the possibility that we can predict with full confidence the reoccurrence of those chains of cause and effect we can sensibly perceive. Instead, he asserts that we can only speculate on the probability of the same cause and effect sequence co-occurring based on our experience of their previous co-occurrences:

> We use our *experience* of the constitution and action of coins in general to assert that heads and tails are equally probable, but we have no right to assert before experience that, as we know nothing of nature, routine, and breach of routine, they are equally probable. In our ignorance we ought to consider before experience that nature may consist of all routines . . . and that all such are equally probable. Which of these constitutions after experience is the most probable must clearly depend on what that experience has been like. (172–73)

Along these lines, Pearson outlines a stricter, more self-critical, self-denying version of positivism, which I will call *probabilistic positivism*, in which no absolute, unchallengeable universal assumptions about nature can be made. Instead, the scientist can only speak of his claims as *likely* to be true. This unique aspect of Pearsonian positivism suggests a central role for statistics and mathematical probability in the characterization of the reliability of scientific knowledge. In fact, Pearson compares scientific trials to drawing either a white or black ball from an urn filled with an unknown mixture of balls of each color. This is a reference to Jakob Bernoulli's famous thought experiment on probability in which he explains how to calculate the probability of unknown future occurrences using past occurrences. Pearson writes:

> We are now in a position to return to our bag of black and white balls. . . . We must assume our "nature bag" to have every possible constitution . . . to do this we suppose an infinitely great number of balls in all. We may then calculate the probability that with each of these constitutions the observed result, say p white balls and q black balls . . . would arise in $p+q$ drawings. (176)

This mathematical method for judging scientific validity was not limited, Pearson believed, to the physical sciences alone. It was also applicable in the biological sciences:

> If we look upon biology as a conceptual description of organic phenomena, then nearly all of the statements we have made with regard to physics will serve as canons for determining the validity of biological ideas. In particular, any biological

> concept will be scientifically valid if it enables us to briefly summarize without internal contradiction any range of our perceptual experience. (395)

In this positivistic biology, the goals of physics merge with the goals of biology. Both disciplines would be interested in describing the widest ranges of phenomena in the briefest possible formulae within the routine of perceptual experience (*Grammar*, 1892, 394). However, scientists would need to avoid assuming that important physical categories like force and motion could be applied similarly to biological phenomena (392).

A brief examination of Karl Pearson's life and work reveals that, prior to his serious engagement with the study of heredity and the biometric debates, there were a few important intellectual commitments that influenced his thinking. First, Pearson was committed to his own, unique brand of probabilistic positivism whose scientific ethic limited the content of knowledge to the perceivable and the reliability of knowledge to the probable. Second, Pearson believed in the importance of mathematics in defining and advancing scientific knowledge. With the help of mathematical probability, he believed that scientists could calculate the degree to which their assertions could be assumed to be likely or unlikely. Finally, he maintained, despite his own positivistic convictions, that mathematical reasoning unaccompanied by observation or experiment could be used as a legitimate first step to developing theories about nature. These commitments—to mathematics, probability, and positivism—help explain why, in the latter half of the last decade of the nineteenth century, Pearson became a central figure in the biometrical campaign to create a science of heredity which elevated mathematics and observation above all other methods for prosecuting scientific investigation of the subject.

Frank Raphael Weldon

Despite Pearson's convictions in *The Grammar of Science* that biology might someday follow in the footsteps of physics to become a mathematical science, he was not initially receptive to the idea of a statistical approach to the study of heredity and natural variation. In a public review of Galton's *Natural Inheritance*, given on March 11, 1889, at a meeting of the Men and Women's Club, for example, Pearson commented that the book was interesting and important, but that he saw "considerable danger in applying the methods of exact science to problems in descriptive science, whether they be problems of heredity or political economy" (qtd. in Porter 213). Pearson's initial resistance to Galton's probabilistic/statistical method raises the question: "How did Pearson move from a position of skeptical ambivalence of Galton's method to being one of its most ardent supporters?" The answer involves a combination of a movement in Pearson's own interests towards statistics, and his contact and collaboration with the biologist Frank Raphael Weldon (Magnello 44).

Weldon (1860–1906) was a zoologist whose primary interest was in the evolution of marine fauna. He had graduated from Cambridge in 1881 and was made a lecturer in invertebrate morphology there in 1884. During the earliest period of his research, he worked on developing Darwin's theory of evolution through the study of morphology and embryology. One of the most important, unanswered questions about heredity in the last decades of the nineteenth century was how physical variations occurred and how they accumulated to create diverging physical characteristics in populations of organisms. Some natural researchers believed that variation emerged suddenly as "leaps" in the physiology of an organism. This model, known as *discontinuous variation*, was supported by Francis Galton, who pointed to the occasional emergence of "sports"—or new forms of organisms—as proof that nature made leaps in her design. He argued that these leaps to a new type avoided the problem faced by Darwin's continuous model of inheritance that small changes would likely be blended away over time as animals bred, making it unlikely if not impossible that a particular variation could ever emerge as a stable trait in a population.

In his early morphological and embryological work, Weldon attempted but failed to find evidence supporting Darwin's famous dictum, "*natura non facit saltum*": "nature does not make leaps" (*Origin* 158). As a consequence, in 1889, after reading Galton's *Natural Inheritance*, he decided to approach the issue from a statistical perspective. In May of 1889, he and his wife Florence collected measurements of the length of the carapaces, the exoskeleton covering the first main body segment, of adult female shrimp (*Cragon vulgaris*) and several other measurements[1] at three locations on England's southern, eastern, and western coasts. [2] As a supporter of Darwinian continuous variation, Weldon expected that no symmetry would be found in physical features of these populations because continuous variation would move them towards increasing heterogeneity. After corresponding with Galton for help with the mathematics, however, Weldon confirmed that Galton's argument for the regular symmetrical distribution of traits was not only true for each of the distinct shrimp populations, but also across the aggregate of populations. This meant that these populations were homogeneous, a characteristic of a population influenced by discontinuous variation.

In 1890, he published his findings in the article, "The Variations Occurring in Certain Decapod Crustaceans. I. *Cragon vulgaris*," in the *Proceedings of the Royal Society*. In the paper he explicitly recognizes that the data fit Galton's discontinuous model of inheritance:

> In his recent work on Heredity, Mr. Galton predicted that selection would not have the effect of altering the law which expresses the frequency of occurrence of deviations from the average: so he expected that the frequency, with which deviations from the average size of an organ occurred, to obey the law of error in all cases, whether the animals observed were under the action of natural selection or not. The results of the observations here described are such as to fully justify Mr. Galton's prediction. (446)

Unable to find confirmation of Darwin's theory of discontinuous variation in shrimp, Weldon decided to expand his investigation to other crustaceans and to examine populations that were geographically further separated from one another. Towards this end he examined shore crabs (*Carcinus moenas*), collecting twenty-three measurements from two sets of 1000 female crabs, the first from Plymouth sound, and the second set from the Bay of Naples. Of the twenty-three characters measured, all but one distributed normally according to the law of error. However, this one exception provided an interesting case (Magnello 50).

What Weldon found was that his data from the frontal measurements of crab shells seemed bimodal, or double humped, rather than the expected single hump of a standard, bell-shaped distribution. This discovery excited Weldon because it suggested that he had either identified a population in which a leap had been made to a new form that was being spread throughout the population by natural selection, or that he was witnessing the confluence of two distinct species of shore crab, which had heretofore not been identified. He shared these possibilities with Galton, to whom he wrote on November 27, 1892, "Apart from any arithmetical analysis, I tried to draw inside it two 'Curves of Error' whose sum might represent the observed distribution fairly well. . . . Either Naples is the meeting point of two distinct races or a "sport" is in the process of establishment" (qtd. in Magnello 51).

At the same time Weldon wrote to Galton about his findings, he also wrote to Pearson telling him about his success in mathematically identifying a bimodal distribution in his data for shore crabs. Pearson, who had been preoccupied with the mathematics of frequency distributions, thought that Weldon's example might profitably be studied. He suggested that a more precise result could be found using a process called "curve fitting," in which the original values of the data were re-evaluated using multiple, smaller curves that more precisely followed its referent contours. Pearson worked on the data for all of the summer of 1893. According to his results, the two humps in the curve divided nicely into two, normally distributed curves, which he believed revealed the existence of variation as a result of natural selection in the population (52).

From the results of these efforts, Pearson became excited about the possibility of a purely mathematical approach to the issue of evolutionary variation. His excitement is clear in his notes for his Gresham lectures in 1893, in which he writes:

> We are living in an essentially critical period of science, when more exact methods and more sound logic . . . [are] replacing the old 'scientific gospel.' . . . For the first time in the history of biology, there is a chance of the science of life becoming an exact, a mathematical science. Men are approaching the question of heredity and evolution from a new standpoint. (qtd. in Porter 237–38)

In his Gresham lectures of November 21st and 22nd, 1893, on probability and the normal curve, Pearson used Weldon's work to discuss the potential usefulness of statistics to (1) solve the problems of evolution, (2) dispute the ubiquity of the normal distribution in nature, and (3) discuss his own methods of curve fitting to deal with skewed and bimodal curves. In the text of his November 21st lecture, he writes: "Symmetry is by no means universally the case. . . . The keynote to the most interesting and valuable problems in evolution lies in the non-symmetry of the frequency curves corresponding to the measurements of special organs in animals" (qtd. in Magnello 54). After this statement, he introduces curve fitting for symmetrical and asymmetrical distributions.

In Pearson's writings and his Gresham lectures, it is possible to witness the confluence of different strands of his research from graphical calculation to scientific theory, and the influence of the statistical work in biology on the direction of these combined interests. In Weldon's work, Pearson found the data he needed to illustrate (and, perhaps, even to provide exigency for) his theoretical work with non-normal and asymmetrical curves. It also provided him with an opportunity to make claims about biological processes based on statistical modeling. Though Pearson by no means spent the majority of his time in the latter part of the decade working exclusively on developing biometric theories, he did devote a substantial amount of it to creating mathematical models to describe variation, evolution, and heredity.

Though Pearson was enthusiastic that a statistical approach might lead to the unraveling of the mysteries of heredity and variation, Weldon was a bit more cautious about exclusive reliance on statistics to understand these phenomena. In fact, he complained to Galton about Pearson's singular focus on the issue. In an 1895 letter, he wrote: "Pearson seems to me to reason loosely, and not to take any care to understand his data. . . . He does not see that it is a matter of experiment and not for a priori reasoning at all!" (Weldon to Galton February 11, 1895). Further, Weldon was not supportive of Pearson's interest in skewed curves, which he believed offered the false impression that the crab populations he had investigated were changing rapidly, when in fact he believed they were changing slowly in accordance with Darwin's doctrine of variation (Weldon to Galton Januaru 27, 1895).

An examination of Weldon's early biometric investigations suggests that his work had the hallmarks of pre-paradigmatic science. Though he heavily favored Darwin's theories on selection and variation, he also accepted the possibility of Galton's position on discontinuous variation. In addition, Weldon was convinced that a statistical approach could provide insights into variation and evolution. However, in his critique of Pearson's work, it is clear that he supported experimentation as well.

William Bateson

William Bateson (1861–1926), like Weldon, was a Cambridge trained biologist. However, unlike Weldon, he gravitated towards the idea that discontinuous variation was the driving force behind natural variation. In 1891, Bateson began to seriously develop his theory of discontinuous variation, articulating it in an article he co-wrote with his sister Anna Bateson, entitled "On the Variations in Floral Symmetry of Certain Plants Having Irregular Corollas," in the *Journal of the Linnaean Society*. In this article, the authors are careful not to dismiss the reality and importance of continuous variation; however, they do question the assumption that continuous variation is the only or even primary source for variation in natural forms (Punett, *Scientific Papers* 158–59).

In 1892, the same year that Weldon gathered and examined his data on crabs, Bateson published, "Some Cases of Variation in Secondary Sexual Characteristics Statistically Examined," in the *Proceedings of the Zoological Society* in which he presented statistical evidence for the bimodal distribution of characters in the forceps on earwigs (*Forficula auricularia*) and in the horns of stag beetles (*Xylotrupes gideon*). This work contains results that are remarkably similar to those published by Weldon in the following year. Like Weldon, Bateson takes the position that a bimodal distribution is present in his data, and this distribution suggests discontinuous variation:

> In the common language of naturalists, the facts of this case [the forceps on earwigs] suggest that there is, for some wholly unknown reason, a dimorphism among males of these earwigs, maintained though all live together. . . . For the present we are content to recognize that in this case of the earwig there is evidence of a definite and partially discontinuous variation in respect of a secondary sexual character (qtd. in Punett, *Scientific Papers* 197).

Despite the philosophical differences between Weldon and Bateson about whether discontinuous or continuous variation was the predominate force in evolution, Pearson saw Weldon's crab curves and Bateson's earwig curves as statistically similar, and included both as examples of bimodal curves in the printed abstract of his paper, "Mathematical Contributions to the Theory of Evolution" (1893). He writes, "Such abnormal curves arise particularly in biological measurements; they have been found by Professor Weldon, for the measurement of a particular organ in crabs, by Mr. Thompson for prawns, by Mr. Bateson for earwigs" (Pearson, "Mathematical Contributions,"

329–30).[3] What is interesting about Pearson's early comments on Bateson's work is that he sees no significant difference between the conclusions of the two researchers.

Though their data might have distributed in the same way, Bateson's focus, which is clear in his major work, *Materials for the Study of Variation* (1894), is on the importance of discontinuous variation. In this work, Bateson catalogues facts about variation in specific organisms and speculates about the nature of variation based on these collections of facts. He argues that the evidence supports his claims that, (1) discontinuous variation exists,(2) this variation has its source within the organism, and (3) this type of variation is not a product of natural selection.[4] He writes, "The discontinuity, of which species is an expression, has its origin not in the environment nor in any phenomenon of adaptation, but in the intrinsic nature of organisms themselves, manifested in the original discontinuity of variation" (567).

Despite his attention to discontinuous variation, Bateson does not completely reject the possibility that continuous variation, in the form of blended inheritance, may play some limited role in variation. In the concluding remarks of *Materials for the Study of Variation*, he recognizes the existence of blended inheritance and its contribution to variation when he writes:

> Though it is obvious that there are certain classes of characters that are often evenly blended in the offspring, it is equally certain that there are others which are not. In all this we are only able to quote case against case. No one has found general expressions differentiating the two classes of characters. (573)

After arguing for the existence of discontinuous variation and its likely role in speciation, Bateson concludes the text with a discussion about the methods whereby these issues might be further explored and clarified. In this discussion he shows an equanimity towards methodological approaches, arguing in favor of a continued program of gathering statistical data while at the same time asserting that this program should not be purely observational but should also rely on experimental breeding to direct or limit the scope of the data set: "The only way in which we may hope to get the truth is by the organization of systematic experiments in breeding, a class of research that calls perhaps for more patience and more resources than any other form of biological inquiry" (574).

By examining the philosophical positions of Bateson, Weldon, and Pearson in the opening years of the final decade of the nineteenth century, this section offers important context for understanding both the individual commitments of the researchers as well as the initial relationship of those commitments to one another. Although there are clear differences within this group, particularly between Bateson and Weldon about the nature of variation, what stands out in comparison is the recognition that there are a number of legitimate and alternative ways of thinking about the phenomenon of variation. Though Weldon favors Darwin's theory of continuous variation, he remains open to the possibility that the normal distribution of his shrimp and crab data might represent stable, homogeneous populations of organisms and the presence of sports produced by discontinuous variation. Similarly, though Bateson does not accept the primacy of continuous variation in creating natural variation, he does not completely dismiss its influence on the process of organic change. Finally, Pearson, Weldon, and Bateson all believe in the value of Galton's application of statistical probability to the study of heredity and endorse it as a method for studying nature. Additionally, both Bateson and Weldon support a mixed regimen of data gathering that involved both observation and experimentation.

The contextual information from this section, which suggests that an environment of theoretical and methodological broadmindedness existed to some extent between these researchers, provides a benchmark for assessing the changes that take place in the relationship between these two groups in the final years of the nineteenth century. With the reemergence of Mendel's work, in conjunction with Pearson's efforts to establish a new statistical research program to investigate variation, evolution, and heredity, the tolerance for coexisting programs of statistical research would become strained and eventually break. In this environment of competition for paradigm formation and legitimacy, values as distinct positions will be transformed into values as virtues.

Value Hierarchies and the Transformation of Mathematics to Virtue

In the first section of this chapter, the concept of *value* was viewed as *value as division.* From this position it was possible to discuss the variety of theoretical and methodological characteristics of Pearson,

Weldon, and Bateson's philosophies in the pre-paradigmatic period, before the Mendelian-biometric debate, while avoiding the presumption that these values were considered virtues. This second section, however, investigates paradigm formation, a process which involves the transformation of *values as divisions* into *values as virtues* through the creation of value hierarchies. As such, it requires a new understanding of *value* that emphasizes the term's more traditional sense as *virtue*.

In addition to having a limited acceptance in the universal audience, values also can be differentiated from truths, facts, and presumptions because they are often maintained as part of a hierarchy, a larger system in which they are ranked in continuous gradations of importance. According to Perelman and Olbrechts-Tyteca, hierarchical systems can either be unreflective and idiosyncratic, self-reflexive and based on a single guiding principle, or an interconnected network of principles which informs the position of the values in the hierarchy (80). The latter type hierarchy, which relies on an interconnected network, they explain, is the kind most commonly represented in argument:

> We shall see that values are generally considered to be interconnected. . . . Value hierarchies are, no doubt, more important to the structure of an argument than actual values. Most values are indeed shared by a great number of audiences, and a particular audience is characterized less by which values it accepts than by the way it grades them (81).

As we have seen, the values at the pre-paradigmatic stage of science were all shared or accepted in varying degrees amongst Pearson, Weldon, and Bateson, and were not used in the intellectual conversation about variation in any meaningful way to articulate or argue for a primacy of a particular position. However, with the contemplation of paradigm formation, a need arises to create an identity to draw adherents and protect theoretical and methodological boundaries. As Perelman and Olbrechts-Tyteca suggest, this process can be carried out through the construction of value hierarchies: "When a speaker wants to establish value hierarchies or to intensify the adherence they gain, he may consolidate them by connecting them with other value hierarchies" (83).

In the following examination of the development of the biometric movement, we will see that Weldon and Pearson began to assemble in the very last years of the nineteenth century a series of value hierarchies to define their approach and, after the turn of the century, to establish its superiority over the Mendelians. They accomplished theoretical distinction by gathering under their program the values of discontinuous variation, the importance of effect, and the importance of mathematical approaches to exploring biological phenomena. By hierarchizing these values, the biometricians created a set of virtues for doing science that defined their program of investigation and cast the Mendelians as adherents of a *distinct*, but *not virtuous*, set of values, including the centrality of discontinuous variation, the importance of cause, and the importance of experimental approaches. For the most part, these distinctions stuck and were not challenged by the Mendelians because of the theoretical commitments imposed by Mendel's work. However, in the case of mathematics, the Mendelians fought back not by trying to establish that mathematics was an organizing principle in their hierarchies of value, but by trying to show that establishing a methodological hierarchy which privileged mathematics over other methods was unnecessarily restrictive.

Formation of the Biometric Paradigm

The rift that developed between Bateson and the biometricians in the middle of the last decade of nineteenth century was precipitated by events set into motion by Francis Galton. Though Galton was largely responsible for Pearson's, Weldon's, and Bateson's interest in a statistical approach to the study of heredity, his efforts to bring together the spectrum of values supported by these researchers within a single, institutionally supported committee proved to be disastrous. While well-intentioned and reflective of the general spirit of theoretical and methodological tolerance of the time, by introducing real stakes into the pre-paradigmatic environment—specifically the possibility of a funded scientific program of research—Galton created the conditions for agendas to develop with a view towards consolidating power around particular ontological and epistemological frameworks.

Impressed with the results of Weldon's statistical studies on shrimp and crabs, and encouraged that the work would lead to further breakthroughs, Galton threw his considerable clout and fortune behind the formation of a committee to support statistical research on plants and animals. On January 18, 1894, with Galton as the committee chair and Frank Weldon as secretary, the Committee for Conducting Statistical Inquiries into the Measurable Characteristics of Plants and

Animals was founded under the charter of the Royal Society for the purpose of funding statistical research regarding questions of evolution and inheritance (Gillham 299).

Initially, neither Pearson nor Bateson were members of the committee. However, because of his work with Weldon and his zealous support of a mathematical approach to evolution, Pearson was added at the end of 1896. In addition to adding Pearson, Galton also added Bateson both to settle a row between Bateson and Weldon over the latter's paper, "Attempt to Measure the Death Rate Due to the Selective Destruction of *Carcinus Mænus* with Respect to a Particular Dimension," and to promote an experimental breeding program of which Galton was in favor. [5] In the next year, the expanded committee was renamed "The Evolutionary Committee of the Royal Society" at Galton's request (Gillham 300).

Despite Galton's good intentions to bring traditional biologists, mathematicians, and quantitatively inclined biologists together, the resulting combination was an infelicitous mix. Though there was general civility between Weldon and Bateson—as evidenced by their cooperation in advertising and working out the details of the experimental breeding program[6]—Weldon and Pearson still felt an underlying discomfort on the committee because they believed their enthusiasm for statistical investigation was not shared by most of its members. This discomfort came to a head at the end of 1899. In November of that year, Weldon wrote to Galton:

> We held our meeting yesterday. There were present F. Darwin, Bateson, Heape, E. Clarke, and myself. . . . The discussion seemed to show more than ever the futility of the committee. . . . The most dangerous thing about our discussion seemed to me the deliberate antithesis which these men sought to establish between what they called "Natural History" on the one hand, and any sort of statistical enquiry, leading to numerical results on the other.
>
> Several members of the Committee, of whom Bateson is the leader, have the idea of such an antithesis, and are strongly influenced by it. However sincere and able such men may be, I think they do harm, and I think it would be a great misfortune if a Committee of the Royal Society were to officially adopt their view.

> I cannot go on acting as secretary of this committee. Will you advise me what to do? (Weldon to Galton 28 Nov. 1899)

These lines are evidence that what had previous been distinct but not incompatible values had suddenly begun to aggregate into value hierarchies of competing virtues. Weldon recognizes these hierarchies in the development of what he describes as "deliberate antitheses" between "Natural History" and "statistical inquiry," which he claimed Bateson and other researchers began to make. The stakes of these divisions are an element of concern for Weldon, who voices his fears that the Committee of the Royal Society may accept Bateson's position as the official view, thereby elevating his particular hierarchy of values closer to the status of paradigm.

The final question in the letter is a veiled query to Galton about what he believes the fate of the committee should be. If he agrees to resign in protest along with Pearson and Weldon over the activities of the committee, Weldon, who is secretary and tasked with recommending to the Royal Society the reinstatement of the committee's charter, can make a case for the disbandment of the committee. This would require the remaining members to petition for the creation of a new committee. This dramatic option would either block the efforts of the opposition to gain traction for their platform altogether or, at the very least, make it much more difficult for them to receive the official sanction.

On January 25, 1900, Galton, along with Pearson and Weldon, exercise this option, leaving the committee's fate in the hands of the remaining participants. Through some shrewd political wrangling, however, Bateson collected enough support to keep the committee from folding. As a result, the committee was completely dominated by Bateson and his supporters, much to the chagrin of Pearson, who later wrote that their "capture of the committee was skillful and entirely successful" (qtd. in Gillham 307).

Frustrated by Bateson's consolidation of institutional power in the evolutionary committee, and fearful that his platform of investigating variation, evolution, and heredity would be further marginalized, Weldon set out to establish an alternative institutional presence supporting descriptive statistical investigations of biological phenomena. With Pearson's firm commitment to his scientific agenda and Galton's moral and financial support, the journal *Biometrika* was founded in 1901. Its purpose, according to Pearson's opening editorial, "The Scope of

Biometrika," was to "serve as a means not only of collecting under one title biological data of a kind not systematically collected or published in any other periodical, but also of spreading a knowledge of such statistical theory as may be requisite for their scientific treatment" (1).

Although the opening lines of Pearson's editorial seem to represent some common-sense goals of gathering data and developing a statistical approach to biological questions, further reading reveals that *Biometrika* was also intended to support a more radical platform for the development of a new mathematical biology, guided exclusively in its investigations of evolution and heredity, by the principles and practices of mathematics. In the second section of the editorial entitled, "The Spirit of *Biometrika*," Pearson expresses his views that biometrics must be focused solely on a mathematical solution to the questions of evolution, variation, and heredity:

> Whatever views we hold on selection, inheritance, or fertility, we must ultimately turn to the mathematics of large numbers, to the theory of mass phenomena, to interpret safely our observations. As we cannot follow the growth of nations without statistics of birth, death, duration of life, marriage and fertility, so it is impossible to follow the changes of any type of life without its vital statistics. The evolutionist has to become in the widest sense of the words a registrar-general for all forms of life. (3)

In his mission to create an exclusively statistical domain in the study of evolution, variation, and heredity, Pearson hoped to reach two different audiences: biologists and mathematicians. He makes the dual nature of his audience clear in his idealized description of the statistical scientific enterprise:

> The biologist may find in our pages algebraic analysis which may repel him. We would still ask his attention for the general conclusions and for the formulae reached by the mathematician. The biologist will find that they frequently suggest observations and experiments which he alone is in position to undertake satisfactorily. We shall aid the more arithmetical part of his work by diagrams and numerical tables wherever it seems possible. In this manner we hope that *Biometrika* will provide for both branches of science; that it will not only publish valuable biometric and statistical researches, but serve as a

> storehouse of unsolved problems for both unemployed biologist and mathematician. (5–6)

As in his earlier days working with physical mathematics, this characterization of the relationship between the mathematician and the scientist suggests a hierarchy with the mathematician at the top. In this scheme, the mathematician's job is to come up with formulae based on an assumed set of parameters and to reach general conclusions, which then suggest the type of data to be collected and the subject and manner of experimentation to be carried out by the biologist. According to this biometric vision, science becomes the process of developing and testing mathematical theory through the collection of data. This focus highlights the importance of effect and its measurement as the fundamental principle guiding scientific investigation, organizing the hierarchy of values to place mathematical analysis and description at the vanguard of the scientific process.

Although *Biometrika* had been founded as a safe place for Weldon and Pearson to present their statistical-biological work, it was not the only outlet for Pearson's efforts to establish the value hierarchies in his platform for a mathematical biology. In fact, Pearson had been publishing mathematical papers based on biological examples in the series "Mathematical Contributions to the Theory of Evolution" in the *Philosophical Transactions of the Royal Society* since 1893, and continued to publish there until 1904.

At first, the titles of Pearson's papers were either descriptions of new statistical tools or discussions of new statistical procedures for analyzing the problems presented by biological data. In 1895, however, his work began to change. It moved from developing mathematical methods for describing the distribution of character traits in a population to developing a theory of heredity based on mathematical principles. Emblematic of these efforts was Pearson's paper, "On the Principle of Homotyposis and its Relation to Heredity, to the Variability of the Individual, and to That of the Race," in which he proposes a detailed model of variation and heredity that supports the three major values of the biometric platform: the virtues of continuous variation over discontinuous, the virtue of investigating effect over cause, and the virtue of a quantitative observational approach over an experimental and cytological one.

Homotyposis and Heredity

Charles Darwin believed that variation within a population was the key to its flourishing. When competition for resources within and between populations became fierce, highly varied groups of organisms would have a better chance of surviving and producing offspring because there was a greater likelihood that at least some of their members would have variations that would allow them to exploit resources their competitors could not (Darwin, *The Origin* 93). Because continual diversification was so central to the success of organisms, Darwin believed that natural populations were constantly diversifying in a variety of directions through the random appearance of variation. Although diversification was theoretically central to Darwin's explanation of speciation, it posed a challenge for Pearson who wanted to develop reliable quantitative descriptions of evolution. In order to mathematically study the action of evolution, Pearson needed to identify a constant value for heritable variation. If he could establish such a value, he could determine whether and in what ways selective forces were transforming a population.

To overcome the obstacles to standardization inherent in Darwin's model, Pearson turned to the law of error and particularly to Francis Galton's mathematical concept of correlation. In his "Homotyposis" paper, he proposes a theory of variation which sets the value of heredity within limits and explains its fixity as a consequence of the general similarity between organisms in a species and the propensity of heredity to follow the laws of probability. In the opening of his paper, Pearson lays out these goals in questions he hopes to answer: (1) "What is the ratio of individual to racial variability?", and (2) "How is the variability in the individual related to inheritance in the race?" (286).

To answer the first question—"What is the ratio of individual to racial variability?"—Pearson turns to the concept of *homotypes* or *undifferentiated like organs.* He explains that in nature individual organisms produce organs such as leaves, scales, or blood cells which tend to be similar to one another yet vary slightly in character. He calls these organs homotypes. Reasoning analogically at first, he advances the idea that the production of undifferentiated like organs by an individual organism is similar to the production of offspring. Undifferentiated like organs, such as leaves, vary just as children differ from their parents and from each other, but like children, they share an affinity because they are generated from a common source. He argues that the degree of this affinity of relatedness could be measured using Galton's mathematical method of correlation:

> Just as we can find by the methods [of correlation] already discussed in earlier memoirs . . . the degree of correlation between brothers and the variability of an array of brothers due to the same parentage so we can determine the correlation . . . between undifferentiated like organs of the individual and the degree of variability within the individual. (287)

In the next lines of the paper, Pearson extends this analogy into a causal argument by maintaining that sperm and ovum are undifferentiated like organs. He argues that, if sperm and ovum can be classified as undifferentiated like organs, it follows that the value of correlation for offspring should be the same as the value of the correlation for undifferentiated like organs:

> But turning to the process of reproduction, the offspring depend on the parental germs, and it would thus seem that the degree of resemblance between offspring must depend on the variability of the sperm cells and the ova which may each be fairly considered as "undifferentiated like organs." (287–88)

Following this hypothesis, Pearson explains that if this relationship can be established the study of heredity can be greatly simplified. Variation would no longer be an idiosyncratic character of the individual but a general characteristic of the production of undifferentiated like organs, a uniform process across individuals. This uniformity would allow the correlation between undifferentiated like organs to stand as a constant value for heredity:

> Now the reader will perceive at once that if we can throw back the resemblance of offspring of the same parents upon the resemblance between the undifferentiated like organs of the individual, we shall have largely simplified the whole problem of inheritance . . . If this view be correct, variability is not a peculiarity of sexual reproduction, it is something peculiar to the reproduction of undifferentiated like organs in the individual, and the problems of heredity must largely turn on how the resemblance between such organs is modified. (288)

In the rest of the paper, Pearson endeavors to make the case for a hereditary constant by establishing the correlation value of undifferentiated like organs and by comparing these to the correlation values between offspring and parents. Pearson begins on the assumption that the correlation value—the mathematical degree of relatedness—for offspring, *R*, is the same as the fraternal correlation calculated using a modified version of Galton's Law of Ancestral Heredity *R* = .4. This strategy for argument suggests that Pearson believed that the majority of his audience accepted the Law of Ancestral Heredity as a fact. Therefore, they would be willing to grant that if he could show a correspondence between this value and the correlation value of undifferentiated like organs, he could establish that variation in reproductive organs was responsible for the variation in offspring.

In section III of the paper, "On the Variability and Correlation of Undifferentiated Like Organs in the Individual," Pearson calculates the correlation values of undifferentiated like organs in a variety of plants. His choice of plants seems to reflect his need for easily accessible subjects that produced, what to his mind were unquestionably undifferentiated like organs, physical features that were numerous and only slightly varied from one to another. These included physical features such as leaves, seeds, and gills on mushrooms. The relative similarity of these organs was important because it maintained Darwin's position that variation was generally slight.

In addition to selecting his subjects on the basis of their preponderance of undifferentiated like organs, Pearson also chose organs to measure that were substantially different from one another. Though this criterion has little impact on the investigation of correlation of organs in individuals, it was important to his position that his general correlation value held reliably across different types of organisms:

> What I have endeavored to do is to take as wide a range of different organs as possible in different types of life and trust to the bulk of my statistics to give me a substantially accurate value of ρ [correlation of undifferentiated like organs] to compare to the values of R [fraternal correlation] we have determined on other occasions. (292)

True to his word that he provide evidence concerning, "as wide a range of organs as possible," Pearson provides correlation values from undifferentiated like organs such as leaf veins and pinnea, mushroom gills, and Shirley poppy seeds. Using leaf veins as an example, he explains his statistical method for calculating correlation. In conducting this type of analysis, twenty-six leaves were gathered per tree from around one hundred trees in two or more distinct regions (for example, Bukinghamshire, Dorsetshire, and Monmouthshire) at various positions and heights on each tree. The purpose of gathering leaves from different regions of different trees in a variety of areas was to minimize the possibility that the values of homotyposis might be skewed by factors of growth, climate, soil, etc.

Once collected, the number of veins in each leaf were counted and then compared to the number of veins on every other leaf on a given tree. The number of times a particular pairing occurred (for example, the number of times a leaf with three veins was paired with a leaf with nine veins) was listed on a chart, generating a total of 325 entries (½(26x25)).[7] In order to maintain symmetry, Pearson also created a second table in which the order of comparison was switched. This resulted in a total of 650 (325x2) entries for each tree, and 65,000 for one hundred trees. Though Pearson does not offer a detailed visual of his counting of leaf veins, his method is illustrated in his table describing the comparisons of the number of leaf pinnea on ash trees, which reports the measurements from over one hundred trees in the area of Buckinghamshire (see Table 5).

Pearson's charts reveal graphically the range of pairings and the distribution of their frequency. These calculations allowed him to calculate the *mean correlation* and *standard deviation* in the distribution of leaves. Using the calculation of *correlation*, he hoped to show that that the value for homotyposis was the same as the value for fraternal correlation, thereby proving that homotyposis was the source of heritable variation. With the values for *standard deviation*, he endeavored to prove: (1) that the extent of variation in the individual was less than the extent of variation in the race, and (2) that the difference between the extent of variation in the race and the individual was very small. If the first conclusion from standard deviation was shown to be true, then Pearson would have evidence challenging Darwin's notion that individuals are constantly diversifying away from the specific mean. If the second position could be verified, it would suggest that species have a modicum of genetic stability, making a genetic constant feasible. Pearson articulates these conclusions at the end of his discussion of experimental techniques.

Table 5. Distribution of paired measurements from leaf pinnea on ash trees. Reprinted from Karl Pearson, "Mathematical Contributions to the Theory of Evolution. IX, On the Principle of Homotyposis and Its Relation to Heredity, to the Variability of the Individual, and to that of the Race. Part I. Homotypos in the Vegetable Kingdom," *Philosophical Transactions of the Royal Society of London. Series A, Containing Papers of a Mathematical or Physical Character* 197, Table I p. 297. ©1901. Used by permission of the publisher, The Royal Society.

TABLE I.—(i.) Buckinghamshire Ashes.

Number of Pinnæ on First Leaf. (columns) / *Number of Pinnæ on Second Leaf.* (rows)

	3.	4.	5.	6.	7.	8.	9.	10.	11.	12.	13.	14.	15.	16.	Totals.
3	6	0	6	6	15	0	6	0	33	3	—	—	—	—	75
4	0	0	0	0	0	0	0	0	0	0	—	—	—	—	0
5	6	0	18	16	74	14	155	18	94	3	2	—	—	—	400
6	6	0	16	24	96	25	213	29	113	3	0	—	—	—	525
7	15	0	74	96	716	211	2168	235	1385	34	87	0	4	—	5025
8	0	0	14	25	211	50	732	82	485	17	56	0	3	—	1675
9	6	0	155	213	2168	732	8970	1305	7361	278	756	11	20	—	21975
10	0	0	18	29	235	82	1305	278	1648	69	213	6	17	—	3900
11	33	0	94	113	1385	485	7361	1648	13558	838	2825	57	103	—	28500
12	3	0	3	3	34	17	278	69	838	90	326	19	20	—	1700
13	—	—	2	0	87	56	756	213	2825	326	1896	88	176	—	6425
14	—	—	—	—	0	0	11	6	57	19	88	16	28	—	225
15	—	—	—	—	4	3	20	17	103	20	176	28	54	—	425
16	—	—	—	—	—	—	—	—	—	—	—	—	—	—	0
Totals	75	0	400	525	5025	1675	21975	3900	28500	1700	6425	225	425	0	70850

> That variation in the individual is less than that of the race . . . I shall speak of as *homotyposis*. Thus homotyposis denotes not only the likeness of the homotypes [both individual and racial], but that this likeness has probably definite quantitative limits. If my view be correct, heredity is only a special case of homotyposis . . . and denotes a given degree of variation and a given degree of likeness. (294)

After offering his conclusions and methods, Pearson presents the reader with his data and his calculations of their correlation values and standard deviation. In the objectivist spirit of full disclosure, Pearson presents his data and the results of his calculations in sections four and five of the paper. In these sections the data are carefully arranged such that the most compelling evidence is presented first, and less promising results later. This strategy is a classical rhetorical tactic of emphasis described in the *Rhetorica ad Herennium*: "In the proof and refutation of arguments it is appropriate to adopt and arrangement of the following sort: (1) the strongest arguments should be placed at the beginning and at the end of pleading; (2) those of medium force . . . should be placed in the middle" (18; bk. III, sec.x).

If we compare the value of the results presented in section four "Actual Data" with the value of Galton's fraternal correlation R=.4, a clear arrangement strategy emerges, suggesting that Pearson's ordering of evidence maximizes the reader's attention to the results that support his conclusion, and minimizes their attention to results that challenge them. The first data set, the correlation of leaf pinnae on ash trees, produces a result of .39, a scant .01 difference from Galton's value. In the last data set, the correlation of seed pods in plants of the legume *Cytisus scoparius*, the value of correlation is .4, an exact match to Galton's value. Immediately following the first data set, the values rise slightly to .5, .59, etc. Immediately preceding the last data set they are also slightly higher, with values of .6. In the middle sets, however, there is a more substantial change in value. From the correlation data on the number of segmentations of seed vessels in *Nigella Hispanica* to the number of leaves circling the stem of the plant in *Asperula odorata*, the correlation values dip to between .17 and .189, a substantial difference of .23 and .21 from the target value.

Although Pearson's initial strategy to preserve his credibility may have slipped by the incautious reader, his gambit to maintain an ethos of objectivity becomes more challenging when he has to argue for the compatibility of his correlation values with Galton's across his full data set. The problem Pearson runs into is best understood by appealing to Perelman and Olbrechts-Tyteca's notion of *presence.* According to the authors, arguers endow *presence* on argument features by selecting them to make their case: "By the very fact of selecting certain elements and presenting them to the audience, their importance and pertinence to the discussion are implied. Indeed, such a choice endows these elements with a *presence*" (116).

Absolute presence or full disclosure was an epistemic virtue for maintaining an ethos of scientific objectivity; however, presenting all evidence can have the effect of leaving the scientist in the awkward position of having the evidence contradict his or her position. Pearson finds himself caught between maintaining his scientific ethos and defending the virtues of the biometric program enshrined in his theory of undifferentiated like organs. In the opening lines of the "Summary of Results" section, he recognizes his dilemma:

> In summing up my results and comparing them with those obtained from fraternal correlation by my coworkers and myself, I felt some difficulty. If I made selection of what I considered the best homotypic correlation series and the best fraternal correlation, I might well lay myself open to the charge of selecting statistics with a view to the demonstration of a theoretical law laid down before hand. Accordingly, I determined to include all of my homotypic results except where it was pretty evident that we had . . . an influence exerted by the growth factor. (355)

After his admission, Pearson reports his figures for the correlation of homotypes across the various plants he measured (Table 6). In the column labeled "Race," Pearson lists the location and type of plant that was the source of the homotype measured. In the "Character" column, he offers a description of the homotypic organ. The next column lists the "percentage variation," the numerical value for the degree of relatedness between the individual and the race for the organ listed in the "Character" column. Further right, the "Correlation" column describes the degree of relatedness amongst the sets of organs for the homotypic character. Finally, the mean values for variation and correlation are calculated for all of the races bearing a roman numerical designation.

Table 6. Pearson's summation the values for correlation and variation amongst homotypes in the organisms he has measured. Reprinted from Karl Pearson, "Mathematical Contributions to the Theory of Evolution. IX, On the Principle of Homotyposis and Its Relation to Heredity, to the Variability of the Individual, and to that of the Race. Part I. Homotypos in the Vegetable Kingdom," *Philosophical Transactions of the Royal Society of London. Series A, Containing Papers of a Mathematical or Physical Character* 197, Table XXII p. 356. ©1901. Used by permission of the publisher, The Royal Society.

TABLE XXXII.—General Results for Homotypic Correlation.

Race.	Character.	Percentage variation.	Correlation.	Remarks.
Mushroom, Hampden . . .	Lengths of gills .	50·92	·8607	All these results introduce a correlation due to stages of growth and accordingly are not included in the determination of means.
" " . . .	Breadths of gills .	67·67	·7363	
" " . . .	Lengths and breadths of gills	—	·6275	
Wild Ivy, mixed localities .	Lengths of leaves .	82·73	·5618	
" "	Breadths of leaves	84·56	·5332	
" "	Lengths and breadths of leaves	—	·5157	
(i.) Ceterach, Somersetshire	Lobes on fronds .	77·57	·6311	Said to be largely affected by growth and environment.
(ii.) Hartstongue, Somersetshire	Sori on fronds .	77·64	·6303	
(iii.) Shirley Poppy, Chelsea .	Stigmatic bands .	78·86	·6149	Much selected in transit.
(iv.) English Onion, Hampden	Veins in tunics .	79·18	·6108	Possibly slightly heterogeneous.
(v.) Holly, Dorsetshire . .	Prickles on leaves	80·11	·5985	
(vi.) Spanish Chestnut, mixed	Veins in leaves .	80·65	·5913	Heterogeneous.
(vii.) Beech, Buckinghamshire	Veins in leaves. .	82·17	·5699	
(viii.) *Papaver Rhœas*, Hampden	Stigmatic bands .	82·71	·5620	
(ix.) Mushroom, Hampden .	Gill indices . . .	83·58	·5490	Possibly influenced by individual growth.
(x.) *Papaver Rhœas*, Quantocks	Stigmatic bands .	84·59	·5333	
(xi.) Shirley Poppy, Hampden	Stigmatic bands .	85·18	·5238	
(xii.) Spanish Chestnut, Buckinghamshire	Veins in leaves. .	88·51	·4655	
(xiii.) Broom, Yorkshire . .	Seeds in pods . .	90·96	·4155	
(xiv.) Ash, Monmouthshire .	Leaflets on leaves.	91·44	·4047	
(xv.) *Papaver Rhœas*, Lower Chilterns	Stigmatic bands .	91·66	·3997	All from one field.
(xvi.) Ash, Dorsetshire . . .	Leaflets on leaves.	91·81	·3964	
(xvii.) Ash, Buckinghamshire .	Leaflets on leaves.	92·73	·3743	
(xviii.) Holly, Somersetshire .	Prickles on leaves	93·12	·3548	
(xix.) Wild Ivy, mixed localities	Leaf indices. . .	96·21	·2726	From two localities and possibly slightly influenced by differentiation.
(xx.) *Nigella Hispanica*, Slough	Segments of seed-capsules	98·18	·1899	Differentiation of organs due to position on stem.
(xxi.) *Malva Rotundifolia*, Hampden	Segments of seed-vessels	98·32	·1827	Principally spread from one clump by stolons.
(xxii.) Woodruff, Buckinghamshire	Members of whorls	98·49	·1733	Members really different in morphological origin.
Mean of 22 cases . .	—	87·44	·4570	—

Despite Pearson's recognition that data selection could be conceived of as manipulating the numbers to prove one's hypothesis, he does not use all of his results in calculating the mean correlation and standard deviation for undifferentiated like organs. The measurements of correlation in mushroom gills and ivy leaves that are placed above the horizontal line at the top of the chart and not given roman numerals are not included in his calculations of correlation and variation. A note in the far right column of the chart explains that these values are not included because Pearson believes these measurements introduce a false correlation unduly influenced by "stages of growth" (356). This elimination is acceptable in Pearson's opinion because growth would make the values of the homotypes unstable and, therefore, not comparable, just as it would be improper to draw conclusions about the correlation of stature of male humans by measuring them in seven-year-old boys because their statures would be still in a state of flux.

A close inspection of the correlation values on the chart, however, reveals that Pearson's elimination of ivy leaves seems to be motivated by a desire to fit the data. The correlation values he presents for the length of ivy leaves, their breadths, and the combination of their lengths and breadths are respectively .56, .53, and .51. A glance at the values on the acceptable portions of the chart reveals that there are no less than eight of a total of twenty-two acceptable measures which actually exceed the highest value presented by the ivy, and a total of eleven that are larger than the lowest value in the ivy measurements. If the ivy results based on their value are in fact aberrant, then so are half of the results presented as acceptable.

In addition to irregularities evident in Pearson's eliminating certain values from his calculation of overall correlation, there is also an unusual tendency on his part to admit data that might reasonably be eliminated based on their values falling far outside the normal range. If we take Pearson's own calculation of the mean for the correlations in his research .4570, and the highest correlation value he has accepted into his normal distribution .6311 (for (i.) *Ceterach Sommersetshire* in Table 6), we can assume that he believes that most reasonable values for correlation fall within ± .1741 of the mean (a symmetrical regular distribution of values). If this amount of deviation is the extent of the accepted values for the upper and lower limits, then for the sake of symmetry, any values below .2829 (.4570 - .1741) should be eliminated. A brief inspection of the lower values on Pearson's chart, however, reveals that he has not uniformly applied the same biological scrutiny to the homotypicality of organisms whose values fall below the mean. In fact, four of his correlation values (race types xix to xxii in Table 6), almost 1/5 of the data, fall below this limit, yet they are still included in the final computation. They are maintained despite Pearson's own recognition in the far right column of the table that there might be environmental, morphological, or physiological factors affecting these values.

By taking out values that fall above the mean but below the upper limits of the included data, and including data whose values fall below the lower limits, Pearson artificially lowers the mean value of the distribution. Extending the upper limits of the acceptable data to the same degree as the lower limits and including all of the data that is excluded results in an upwards shift in the mean value of correlation. By including all the data within the new acceptable range, the mean value of correlation moves from .4570, which already differs from his original hypothesis of .4, to .4826, an even greater difference.

In addition to the randomness of Pearson's choices of what data to cut and what data to maintain, his manipulation of results to support his conclusions is evidence by his reevaluation of the fraternal correlation to better suit his hypothesis. In the beginning of the paper, Pearson clearly states that the value of the fraternal correlation is .4 (295). In the results section, however, he gives a different value for this correlation: .4479 (358). This new calculation is made from a combination of statistics taken from humans and animals, including the correlations from a variety of features such as the coat color in basset hounds and horses. In comparison, Galton's original calculation for fraternal correlation was based exclusively on statistics comparing the stature of mid-parents—the combined average of the features of male and female parents after the value of the female parent has been converted to be comparable—to their male offspring.

Pearson presents his range sources for calculating fraternal correlation and their correlation values in Table 33, "General Results for Fraternal Correlation" (Table 7). The arrangement of this table is similar to the one in which Pearson presents his calculations for correlation and standard deviation. It contains the columns "Race," "Character," "Correlation," and "Remarks," as well as a separate section at the bottom where the mean value of the fraternal correlation is given. There are, however, a few notable differences. Because the value of fraternal correlation is arrived at by comparing the stature of the mid-parent to the stature of male offspring, gender plays a role in these calculations.

To account for the gender of the parents and offspring being compared, Pearson includes a column marked "Sex" in which either male

Table 7. Pearson's list of sources for calculating fraternal correlation. Reprinted from Karl Pearson, "Mathematical Contributions to the Theory of Evolution. IX, On the Principle of Homotyposis and Its Relation to Heredity, to the Variability of the Individual, and to that of the Race. Part I. Homotypos in the Vegetable Kingdom," *Philosophical Transactions of the Royal Society of London. Series A, Containing Papers of a Mathematical or Physical Character* 197, Table XXIII p. 357. ©1901. Used by permission of the publisher, The Royal Society.

TABLE XXXIII.—General Results for Fraternal Correlation.*

Race.	Sex.	Character.	Source of material.	No. of cases.	Reduced by	Corre-lation.	Remarks.
(i.) Daphnia	♀ & ♀	Length of spine	ERNEST WARREN	330	K. PEARSON	·6934	Probably much too high, owing to heterogeneity introduced by the selection of a few mothers only.
(ii.) Horse	♀ & ♀	Coat-colour.	WEATHERBY'S Studbooks	1000	K. PEARSON, L. BRAMLEY-MOORE, and A. LEE	·6928	Probably much too high, owing to heterogeneity introduced by the use of comparatively few sires.
(iii.) ,,	♂ & ♂	,,	,, ,,	1000		·6232	
(iv.) ,,	♂ & ♀	,,	,, ,,	1000		·5827	
(v.) Man	♀ & ♀	Forearm.	PEARSON, family data	441	A. LEE	·5424	One pair only from each family.
(vi.) Hound (Basset)	mixed	Coat-colour.	GALTON, from studbook	—	K. PEARSON and A. LEE	·5257	All members of litter without regard to sex.
(vii.) Man	♂ & ♂	Eye-colour	GALTON, family data	1500	K. PEARSON	·5169	All possibly pairs in family taken.
(viii.) ,,	♀ & ♀	Cephalic index	FRANZ BOAS, N. A. Indians	—	C. FAWCETT	·4890	Paternity doubtful.
(ix.) ,,	♂ & ♀	Eye-colour	GALTON, family data	1500	K. PEARSON	·4615	See remark to (vii.).
(x.) ,,	♀ & ♀	,,	,, ,,	1500	,,	·4463	,, ,,
(xi.) ,,	♀ & ♀	Stature	,, ,,	595	,,	·4436	,, ,,
(xii.) ,,	♂ & ♂	,,	,, ,,	605	,,	·3913	,, ,,
(xiii.) ,,	♂ & ♂	Cephalic index	FRANZ BOAS, N. A. Indians	—	C. FAWCETT	·3790	See remark to (viii.).
(xiv.) ,,	♂ & ♀	Stature	GALTON, family data	1181	K. PEARSON	·3754	See remark to (vii.).
(xv.) ,,	♂ & ♀	Cephalic index	FRANZ BOAS, N. A. Indians	—	C. FAWCETT	·3400	See remark to (viii.).
(xvi.) ,,	♀ & ♀	Longevity	Quaker records	1050	M. BEETON	·3323	Reduced below true value by non-selective deaths.
(xvii.) ,,	mixed	Temper	GALTON, family data	1294	K. PEARSON	·3167	Character very indefinite, and difficult to estimate.
(xviii.) ,,	♂ & ♂	Longevity	Peerage records	1000	M. BEETON	·2602	See remark to (xvi.).
(xix.) ,,	♂ & ♀	,,	Quaker records	1947	,,	·1973	,, ,,
—	—	—	—	—	Mean of 19 series	·4479	—

* [Since the above memoir was written I have deduced another exceedingly interesting value for fraternal correlation from the measurements of Prof. C. B. DAVENPORT on the statoblasts of the Bryozoa (*Pectinella Magnifica*, LEIDY). See 'The American Naturalist,' vol. 34, p. 964, 1900. DAVENPORT gives for the standard-deviation of the number of hooks in all statoblasts the value 1·326, and for the average standard deviation of colonies of statoblasts 1·197. If we represent the former by σ, the latter will be $\sigma\sqrt{1-r^2}$, whence I find for the "fraternal correlation" $r = ·4302$, a result in excellent agreement with the mean values we have just found.—July, 1901.]

or female symbols or the word "mixed" appears denoting the gender of the parents and offspring that were measured in the calculations. Assuming that the symbols on the chart are arranged in generational order, the symbol before the ampersand should reflect the gender of the parent, and the symbol after it the gender of the offspring. For example, the symbolic phrase "♀&♀" next to "(i.) Daphnia" would mean that the correlation value for the spinal length of these small planktonic crustaceans was calculated by comparing the spines of adult males with their male offspring.

In addition to the category "Sex," Pearson also includes columns labeled "No. of cases," "Source of Material," and "Reduced by." In this table, the "No. of cases" column is necessary because, unlike table for the deviation and correlation of homotypes, Pearson does not provided a full accounting of the number of organisms measured in the body of the homotyposis paper. By providing the number of cases examined in the table, Pearson gives the reader information for judging the reliability of the conclusions. The more cases examined, presumably the more reliable the value of fraternal of correlation is.

Like the "No. of cases" column, the "Source of material" and "Reduced by" columns are also included to establish the credibility of Pearson's data and results. The "Source of material" column accomplishes this by identifying the provenance of the observational data either by naming the person who gathered it or the text or set of texts in which it was recorded. Similarly, the "Reduced by" column assures the reader that the value for fraternal correlation calculated from the observational data is reliable by identifying the person responsible for its calculation. Before the widespread use of calculating machines, the credentials of the calculator were just as important as that of the observer in vouching for the validity of the mathematical result.

Using the full range of values of fraternal correlation, from planktonic crustaceans to horses to dogs to humans, Pearson calculates a new value for fraternal correlation of .4479, which is higher that the value of .4 posited by Galton. This new figure is reached by admitting figures that are both substantially higher and lower than Galton's value. The highest correlation value is for Daphnia. At .6934, it is almost thirty percent above the value given by Galton. The lowest value of .1973 is twenty percent lower. Because the lower values are less frequent in the sample and quantitatively smaller than the upper values, the mean correlation value is raised by more than four percent.

Though Pearson, in the far right column, recognizes that mitigating factors may be responsible for the extreme variations in the values of fraternal correlation, he doesn't eliminate any of them, raising the fraternal correlation to a number significantly closer to the one for correlation between homotypes described in his first table. As a consequence of the closeness of his homotypic and fraternal correlations, Pearson concludes that the variation measured in offspring is a consequence of the variation in undifferentiated like organs, the sperm and ovum.

> I do not propose to lay great stress on what at first sight might look like a most conclusive equality between the mean values of homotypic and fraternal correlations—within the limits of probable errors .4479 and .4570 are indeed equal. . . . When we associate heredity with sexual reproduction, we are only considering the result of homotyposis (variation and likeness) between individual spermatozoa and between individual ova. (358–59)

Despite efforts to make the strongest possible case for his biometric vision of inheritance, Pearson recognizes the formidable obstacles facing his hereditary theory. In doing so he attempts to inoculate his results from critiques that he had not been mindful of, and from all of the possible causes that might have influenced the outcome of his results:

> I do realize that it is extremely difficult with the complex system of factors influencing living forms to reduce our conditions to that theoretically perfect state in which we shall measure solely the factor we are investigating. . . . In the first place the theoretical conception of undifferentiated like organs is very hard to realize practically. . . . Secondly, the environmental factor comes into play. It is difficult to obtain a hundred individuals with like environment; soil, position with regards to other growths, sunlight, insect life, etc. . . . Thirdly, the difficulty that ensuring all individuals are of the same age or in the same stage of development, is very great. (358)

Though Pearson recognizes these concerns, he also actively rebuts them, drawing on methodological, mathematical principles to support his program of scientific inquiry. Methodologically he points to the extent and variety of his data as a defense against its variability. He argues using the law of large numbers that gathering a lot data from broad sample factors that might skew the result away from the true value will, in the end, cancel each other out, leaving only the true value of individual variation of the undifferentiated like organ:

> If homotyposis had a practically constant value throughout nature, I should only expect this value to be ascertained as a result of the average of many series in which the opposing factors of differentiation, environment, age, stage of growth, etc. may more or less counteract each other. In this manner we may approach to a fair appreciation of the bathmic influence of individuality in the production of undifferentiated like organs. (359) [8]

If there is any doubt in the reader about the Pearson's position on the value of mathematical approach to the study of variation and heredity in comparison to experimental or qualitative observational methods, he dispatches these misgivings at the end of the paper. In the last lines before summarizing the conclusions of his research, he writes:

> Only let us follow the method so clearly indicated by Darwin himself in his 'Cross-Fertilization of Plants;' let us cease to propound hypotheses illustrating them by isolated facts or vague generalities. There are innumerable species in nature ready for us to measure and count. *Sine numero nihil demonstrandum est,* [Without numbers nothing is demonstrated] should nowadays be the motto of every naturalist who desires to propound novel hypotheses with regard to variation or heredity. (362)

In "Homotyposis" Pearson attempts to establish a fixed quantitative value for heredity based on the correlation between undifferentiated like organs. In addition, he provides a methodological exemplar for a new, more rigorous approach to investigating variation, evolution, and heredity using quantitative data and mathematical analysis. By presenting a fixed value for heredity, researchers interested in variation and evolution could turn their attention to quantitatively studying the effects of geographic isolation, fertility, and different schemes of breeding on the frequencies of traits in organic populations. In other words, if Pearson turned out to be right about the value of heredity, his "Homotyposis" paper would serve as the foundation from which a new biometric approach to the study of variation, evolution, and heredity could commence.

Bateson's Critique

While Pearson was certain that his work on homotyposis was revolutionary, some biologists, in particular William Bateson, were skeptical. This final section examines how Bateson, responded to Pearson's efforts in "Homotyposis" to advance his biometric approach. A close textual analysis of Bateson's critiques in the paper, "Heredity, Differentiation, and other Conceptions of Biology: A Consideration of Professor Karl Pearson's Paper 'On the Principle of Homotyposis'" (1902), and in his book, *Mendel's Principles of Heredity: A Defense* (1902), reveals that though Bateson attacked the specifics of Pearson's position, he was very careful not to define his scientific platform in contradistinction to Pearson's. By challenging the specific contents but not the general spirit of Pearson's analysis, Bateson is able to maintain an ethos of reasonability and, at the same time, change the status of mathematics from a *value as virtue* to a *value as difference.* By adopting this approach, Bateson effectively neutralizes a plank in Pearson's biometric platform by making the case that it has no substantive superiority over other modes of investigating nature, while at the same time blunting the effect of any counter-challenge from Pearson that Bateson's work was not sufficiently scientific because it rejected the value of mathematical reasoning to science.

Bateson's efforts to neutralize the biometrician's position on the virtuousness of mathematical argument can be understood with help from the rhetorical concepts of *loci of quantity* and *loci of quality.* In *The New Rhetoric*, Perelman and Olbrechts-Tyteca explain that the *loci of quantity* "affirm that one thing is better than another for quantitative reasons," while the *loci of quality* affirm the value of a thing because it is unique, rare, precarious, or essential (85, 89). Perelman and Olbrechts-Tyteca's loci of quantity and quality refer to arguments at the stasis of *value.* In other words, they describe arguments that use either quantity or quality to make the case that something is better or worse than something else. To make his case that mathematical description and reasoning are the ultimate values or arbiters of truth in biology, Pearson argues from the loci of quality. He makes the case that his mathematical program of biology is *unique* because it is the only way to come to *exact* knowledge about nature. Bateson, as we shall see, opposes Pearson's claim that mathematics is the only pathway towards an exact knowledge of nature by making a *quantitative* appeal: multiple methods are required to develop reliable scientific knowledge about variation and heredity.

At the broadest level of conflict, the debate over the homotyposis paper is about whose approach to the study of variation, evolution, and heredity—Pearson's mathematical exclusivity or Bateson's multimodal inclusivity—offered a more reliable method for understanding these phenomena. However, at the local level there was a more basic dispute between quantity and quality that should not be conflated with the broader argument. The controversy here is at the stasis of *existence*. In this case, the question is not about whether one approach is better than another. Instead, it is about whether Pearson's quantitative measure for homotyposis has any qualitative analog in nature. In this debate, Pearson advocated for the existence of his *quantitative* value for heredity, while Bateson is challenged it on *qualitative* grounds.

By advocating for the reality of his *quantitative* value for heredity at the stasis of existence, Pearson hoped to strengthen his argument from the *loci of quality* at the stasis of value. If his quantitative description of heredity is true, then he can argue that his approach is unique and, therefore, best because it offers the only way to understand variation, evolution, and heredity. In contradistinction, Bateson's challenge of Pearson on *qualitative* grounds at the stasis of existence fortifies his argument from the *loci of quantity* at the stasis of value by demonstrating that multiple approaches need to be used to study variation, evolution, and heredity.

Evidence for Bateson's use of *qualitative* appeals at the stasis of existence to uphold his argument from the *loci of quantity* at the stasis of value can be found in his response to Pearson's homotyposis paper, "Heredity, Differentiation, and other Conceptions of Biology" (1902). In the article, Bateson acknowledges the importance of mathematics as one way of obtaining scientific knowledge but questions whether this one approach is sufficient to develop a clear understanding of the hereditary process. His acknowledgement of the value of biometrics comes in a note appended to the end of the text, in which he writes: "We can feel nothing but admiration for those statistical methods which, as perfected by Professor Pearson, are yielding many useful results not otherwise attainable" (205).

Aside from this brief appreciation of the method, however, Bateson spends the majority of his review pointing out how Pearson's mathematical approach on its own fails to yield exact knowledge about heredity. His primary argument is that, while Pearson is exact about counting outcomes, he lacks the necessary qualitative, biological knowledge to ground his quantification in precise and reliable biological concepts.

In presenting his case against Pearson, Bateson begins with a step-by-step summation of the particular claims that the biometrician is making. He starts by restating Pearson's arguments concerning the assessment of homotypicality in organisms, recalling the basic premises set forward by Pearson: (1) that homotypes, by definition, are a series of organs whose variability has its source in many small random influences, and (2) that the homotypic correlation of undifferentiated like organs in a series can be lowered if variation in an organism has its source in differentiating influences (such as growth rate, environmental factors, etc.).

After laying out the mathematician's claims, Bateson challenges the biometrician's distinctions between *differentiation*, the occurrence of significant variation between two organs which leads to their being distinguished as two different types, and *variation*, the occurrence of slight changes within a homogeneous population of types, which accounts for their distribution according to the law of error. He argues that, from the outset, Pearson's position is in jeopardy because his definitions are tautological and rely on the notions from probability theory rather than on clearly distinguishable features in natural phenomena:

> It is not, however, the difficulty of recognition I would now emphasize, but the fact that between the two phenomena [differentiation and variation] no absolute distinction exists in nature. An "undifferentiated series of like parts" means only a series of like parts which have varied and are varying among themselves but little. A series of highly variable like parts is a series in which differentiation exists or is beginning to exist in a complex and irregular fashion. A "differentiated series of like parts" means a series among which variation is or has become definite or regular. Between these classes there is every shade and degree. No one can say finally where each begins and ends, and, by appropriate selection, we could find homotypic coefficients of any required value. The *average* value of such coefficients taken at random has no significance in nature. ("Heredity, Differentiation" 197)

Because there is no way to reliably classify and thereby separate small from large variations, Bateson argues that Pearson's exact calculations of average value have no validity at all. Though they may be mathematically robust, there is no way of ascertaining whether the features that are plugged into the equations have the characteristics required by the mathematics to make the results meaningful. As a consequence, the average value for correlation has no reliable meaning as a statement about nature.

Pressing further, Bateson gets to the heart of this argument. He proposes to Pearson a hypothetical scenario in which discrimination between variation and differentiation could be made:

> But let us now suppose that we could define differentiation from variation in general, say, as orderly variation. Even so we could not distinguish it unless its order was conspicuous. . . . Does not, then, the presence of orderly differentiation, in various degrees, *compel* us to an analysis of individual instances? In plain language, we shall have to pick and choose our cases, and the value of our coefficient of homotyposis will depend entirely on how we do it. (202)

In these lines, Bateson illustrates the impossibility of a scientific program, such as Pearson's, in which measurement and mathematical reasoning is taken as the only means of developing knowledge about nature. He explains that even if there were a clear division in nature between variation and differentiation, it would still be necessary to look *qualitatively* at individual instances to identify which of these is a case of variation and which is a case of differentiation. As a consequence, the exactness of the science could only come from the cooperation of multiple methodologies, both biological and mathematical, to create an exact science of heredity.

In his final estimation of the shortcomings of Pearson's proposed biometric program of research, Bateson finds it wanting not because it is mathematical, but because its mathematical notions are not sufficiently informed by an exact qualitative knowledge of nature:

> He [Pearson] speaks of the extreme difficulty of determining whether his material is homogenous in respect of the environment, but I miss from his work any deep appreciation of the subtle and evasive quality of differentiation. If anyone would obtain a conception of this difficulty, let him go to any tree

> or large plant and set about pruning it, or better, let him try to choose the shoots for propagation. Until he tries, it seems simple enough; but when he begins, he finds the shoots are of many complexly differing kinds, and unless he has experience pruning or propagating, he will not know which to choose. If he studies the tree attentively, he will soon see that the kinds of shoots are largely definite and, in fact, differentiated. ("Heredity, Differentiation" 202)

In Bateson's critique of Pearson's paper on homotyposis, and in his general critique of biometry, we witness his efforts not to reject mathematics as a way of studying variation, evolution, and heredity, but to change its status from a virtue to a division, one among several important approaches to these phenomena. Just as Darwin, Mendel, and Galton recognized the power and the precision of quantification and mathematical formulae, operations, and principles, Bateson too recognizes their strength and does not attempt to make an issue of Pearson's use of correlation and the law of error to study heredity. What he does take issue with is Pearson's insistence that a mathematical approach is the only or most virtuous route to exact and reliable knowledge about nature. By qualitatively challenging Pearson's quantitative approach to the study of heredity at the stasis of existence, Bateson attempts to strengthen his broader argument from the *loci of quantity* that the study of variation and heredity required more than the single, essential approach advanced by Pearson and the other adherents of biometry.

Conclusion

This chapter has explored *value* as an important rhetorical dimension of the Mendelian-biometric debate. In the first section, *value as division* was used to characterize the status of mathematical arguments about these phenomena in the pre-paradigmatic period after *Natural Inheritance*, and before the reemergence of Mendel's theory of heredity. Evidence from the personal philosophies of Pearson, Weldon, and Bateson during this period revealed that, though each of these researchers had personal preferences for a particular set of values, those preferences in no way inhibited their recognition of other values as legitimate perspectives on variation, evolution, and heredity.

While the first section considered *value as division* in the period of pre-paradigmatic science, the second section investigated the rise of *value as virtue* as a consequence of the emergence of competing programs for the study of variation, evolution, and heredity. With the advent of the Evolutionary Committee, and the prospect of institutionalization, values which had previously been considered divided but complementary positions were transformed into oppositional virtues for the sake of articulating and attacking nascent paradigms. By investigating Karl Pearson's arguments in *Biometrika* and "On the Principle of Homotyposis and its Relation to Heredity," it was possible to witness Pearson's efforts to elevate mathematical method, particularly correlation, and fix a value for heredity using the concept of homotyposis. Bateson's critiques of the paper suggest that his assault on the biometric platform challenges Pearson's elevation of mathematical methods to a virtue, and argues for their return to a more complementary position.

By distinguishing between *value as division* and *value as virtue*, this chapter shows how theoretical distinctions made in rhetoric offer productive ways of identifying and characterizing the emergence of scientific communities. Using these categories, it was possible to describe a shift from pre-paradigmatic science to a nascent period of community formation in which previously complementary positions were transformed into competitive hierarchies of values for the purpose of defining the boundaries of a potential paradigm. That mathematics can participate as a value position in this process suggests that it is more than just a system of rules and procedures that facilitate the prosecution of science. Rather, mathematics can play a central role in the formation of disciplinary identity and be the basis for criticizing not just specific conclusions, but an entire scientific program's ontological and epistemological perspectives on nature. As we shall see in the next chapter, once raised, such a critique is extremely difficult to combat, and may even require challenging assumptions about mathematics fixed at the very core of scientific philosophy.

7 Weightless Elephants on Frictionless Surfaces: The Ethos of Biometry

> *At the present one may say that the mathematical theory of evolution is in a somewhat unfortunate position, too mathematical to interest most biologists and not sufficiently mathematical to interest most mathematicians.*
>
> —J. B. S. Haldane

The triumph of Bateson's Mendelism at the end of the first decade of the twentieth century, and the subsequent rise of mutation theory, pushed Darwin's theory of natural selection and Pearson's biometrical approach to the study of variation, evolution, and heredity into the fringes of genetics research. Testifying to the marginalization of the latter, R.C. Punnett, professor of biology at Cambridge and cofounder of the *Journal of Genetics*, declared:

> Although it is true that most textbooks of genetics open with a chapter on biometry, closer inspection will reveal that this has little connection with the body of work [of geneticists], and that more often than not it is merely belated homage to a once fashionable study. ("Genetics, Mathematics, and Natural Selection" 595)

Though mathematical approaches to the study of variation, evolution, and heredity were at their nadir at the end of the first decade of the twentieth century, they were not headed for complete extinction. Through a mathematical synergy of Mendelian genetics, Darwinism, and biometry, three figures—Ronald Alymer (R.A.) Fisher, Sewall Wright, and John Burdon Sanderson (J.B.S.) Haldane—managed to revive the mathematical program of research in biology, founding what has become known today as population genetics.

As the words of J.B.S. Haldane in the epigraph suggest, however, revitalizing mathematical biology was not a simple affair. On the one hand, it required savvy experts who understood the mathematical complexities of their work and were dedicated to making it accessible to non-specialist biological researchers. On the other hand, it necessitated mediators sympathetic to, and knowledgeable about, conventional biological knowledge and practices who could make the case that a mathematical approach was relevant to and valuable for their work.

This chapter argues that R.A. Fisher possessed both of these characteristics. While it is a truism that Fisher's mathematical work played a fundamental role in establishing that the three fields could be mathematically combined, no investigation of Fisher has commented on the mathematician's rhetorical efforts to persuade non-mathematical biologists that his work was worthy of their attention and to make his mathematical ideas and methods accessible to them. By investigating the texts in which Fisher develops some of his most important arguments and the context in which these arguments developed, this chapter illuminates the importance of rhetoric, in particular ethical appeals, in negotiating an associative relationship between mathematicians and biologists, and its crucial role in the debate about the validity and utility of mathematical arguments in science.

The Shortcomings of Biometrics

The campaign of ethos construction that Fisher was compelled to undertake in his work had roots both in perennial problems and in the special issues that had developed as a consequence of the debate between the Mendelians and the biometricians. In some ways, his rhetorical situation was no different than Galton's or Pearson's. Biologists still had very little training in mathematics and were, therefore, generally unable to understand mathematical arguments, which was a serious impediment to grasping the benefits that might be accrued from using these methods in their own pursuits. In other ways, however, the context was significantly different. Unlike Pearson and Galton, Fisher faced a more hostile audience. They were conscious of the potential shortcomings of mathematical approaches to the study of variation, evolution, and heredity because these failings had been publically aired in the controversy between the biometricians and the Mendelians.

The debate between these two groups had raged over much of the first decade of the nineteenth century. Initially, the contest was a stalemate, with one side then the other seeming to draw the approbation of biological researchers. Even as late as December, 1905, hope was still alive in the biometric camp that they could successfully challenge the conclusions of Bateson and other biologists developing the Mendelian model of inheritance. In a letter concerning a dispute between the Mendelians and the biometricians over the chestnut coat color of horses, for example, Weldon wrote confidently to Pearson:

> I don't see that Bateson has scored at all. He and Hurst are welcome to such help as they can get of my figures [on the inheritance of the chestnut coat color in horses], and in the mean time one can get together some sort of decently made paper. . . . Also, I think the withdrawal [of Charles Chamberlain Hurst's paper on chestnut coat color] will shake the credit of Mendelism in Cambridge, as well as at the R.S. [Royal Society] far more than anything else could have done. (Weldon to Pearson December 23, 1905)

Despite this confidence, by the next year the biometrician's campaign of opposition was over. The quick reversal was not the result of a striking blow to biometrical theories of heredity but because of the untimely death of Weldon in April of 1906.

Weldon's passing cost Pearson a dear friend and his most valuable ally in the promotion of the biometric approach. Though Pearson did publish sporadically on the topics of evolution and heredity after Weldon's death, without this dependable ally in the biological community he could no longer carry on a credible fight. As a consequence, the biometrician's program on the mathematical study of variation, evolution, and heredity generally collapsed, and Pearson redirected his efforts to eugenics and methods of applied statistics (Provine 88).

By the time of the publication of Bateson's book, *Mendel's Principles of Heredity* in 1909, the success of the Mendelian paradigm seemed fairly secure. An examination of twenty-one reviews of his book reveals that thirteen reviewers were positively disposed towards the theories and methods of Mendelian program developed by Bateson and his collaborators for investigating heredity. In most of these reviews, Mendel's theoretical principles were considered a reasonably well-established conceptual framework for thinking about heredity (Appendix B). A reviewer in *Nation* writes, for example, about the general acceptance of the Mendelian ratios, a belief which incorporates the assumptions that heritable characters did not blend, and that traits from both parents were present in their offspring. This was a clear rejection of the biometric position, which argued in favor of blended inheritance and accepted the possibility of exclusive inherence where characters from one parent could be completely absent:

> The Mendelian ratios occasionally fail in practice, but serve nevertheless as a basis for a provisional principle, upon which one may select new material for the obtaining of definite results. Such suggestive work is a distinct contribution to science, and its general results are likely to be wide and very useful. (262)

Though generally successful, there was still skepticism about whether Mendel's theory was the final word on the subject of heredity and variation. At least four of the reviews of *Mendel's Principles*, for example, suggested that more research needed to be done before Bateson and his colleagues could comfortably dismiss objections to their work.[1] Typical of these reviews were comments like the ones expressed by an anonymous reviewer in *Lancet*:

> What is of importance is that these interesting researches should be pursued till a reasonable time has come at which we may ask the Mendelians for light and leading based on their work. That time has not yet arrived. The work is in its infancy, or to be accurate it has just reached its teens, and it is too early to worry about its character or to takes stock of its achievement. (1461)

In addition to their healthy skepticism of the universality of Mendel's theory, many of the more cautiously optimistic reviewers of Bateson's text were also not ready to dismiss out of hand the legitimacy of biometrical methods of analysis. Four of the six reviewers who suggested that more work had to be done to establish the credibility of Mendel's theory specifically mention that the biometrician's statistical approach to studying variation and evolution in large populations deserved more attention. This recognition suggests that, though Pearson largely gave up the fight with the Mendelians after Weldon's death, the ideas of the biometricians, though marginalized, were not altogether forgotten by the scientific establishment. One particularly well-balanced review of the book in the *Manchester Courier* even suggested that the work of the Mendelians and the biometricians were complimentary lines of the same general investigation:

> This controversy appears to us to be unnecessary. . . . The Mendelian is concerned with the physiological interpretation of the facts observed in the *individual* whereas the statistician, or as he is generally called the Biometrician, obtains his conclusions from the statistical study of *masses.* [These two views] form equally important and complimentary parts of a theory of heredity. (10))

Although there were some feelings of general sympathy towards the biometricians, the debate with the Mendelians had also focused attention on the theoretical weaknesses of the biometric approach. In Raymond Pearl's 1915 textbook, *Modes of Research in Genetics*, for example, the one-time student of Pearson's and active proponent of biometry, offers a two point explanation of why the mathematical approach championed by Pearson had fallen out of general favor with biological researchers. The first reason, he argues, was its failure to move beyond description to deal with the issue of causation:

> These [biometric] results constitute no more than a rather precise description of the most superficial external features of the phenomena of heredity. Except only in the simplest of events (and then not directly) a description, however, minute of those events cannot give the slightest real evidence as to their cause. (12)

The second reason was that biometry's lack of consideration for existing biological theories made its conclusions seem irrelevant to the discussions of conventional biologists:

> It has repeatedly been the boast of the biometric writers on this subject that their results were absolutely free from any biological theories. To this some of the more wicked critics have retorted that their results were also quite free from any biological significance. (14)

This second reason closely resembles botanist George Shull's assessment of the debate in the *Botanical Gazette*, when he writes: "These [biometrical] methods are of the greatest importance when rightly used but owing to the almost invariable lack of an equally keen *biological* analysis, the applications of these methods have led to a largely spurious product" (226). These positions—that biometrical approaches lacked (1) a commitment to causation, and (2) a commitment to a practical detailed understanding of biological phenomena—existed, along with the lack of mathematical training amongst biologists, as serious challenges for Fisher in his efforts to persuade his audience to adopt a program of mathematical biology. Though researchers interested in variation, evolution, and heredity seemed to find value in the biometrician's attention to and methods for studying changes in populations, they were not sympathetic to their theoretical commitments to positivistic approaches biological phenomena.

Mendelian Heredity: The Virtue of Knowing Causes

That understanding causation was considered a virtue and aim in the sciences can be traced back as far as Aristotle's *Posterior Analytics*, in which he writes:

> We suppose ourselves to possess unqualified scientific knowledge of a thing, as opposed to knowing it in the accidental way in which the sophist knows, when we think that we know the cause on which that fact depends, as the cause of that fact and no other, and, further, that the fact could not be other than it is. (*The Basic Works* I, ii, 111)

Although Francis Bacon and others challenged the possibility of knowing all of the types of Aristotelian causes (material, formal, efficient, and final), ascertaining efficient and material causation remained a central goal in both the physical and biological sciences into the twentieth century, as evidenced by Pearl's critique of Pearsonian Biometry (*Novum Organum* II, ii 134-35). If Fisher's work had to have this characteristic to be successful, the question is, how was he able to develop a biometrical approach that was more than simply a quantitative description of hereditary outcomes?

Of all of the obstacles faced by Fisher, this issue was perhaps the least difficult to overcome. By the time he began studying at Cambridge in

1909 and formally pursuing his interests in heredity, variation, and evolution, there were already a number of important researchers advancing the possibility that Mendel's model could be used as a basis for mathematical analysis of variation and heredity. By assuming Mendelian parameters as the basis for their mathematical arguments, they also assumed the causal features accepted by Mendelians.

As early as 1902, the statistician Udny Yule had pointed out in "Mendel's Laws and Their Probable Relations to Intra-Racial Heredity," that the biometrical Law of Ancestral Heredity was not incommensurable with Mendelian inheritance. He argued that in a large population of randomly breeding organisms with Mendelian characteristics, those organisms with a *dominant* character would produce offspring in a pattern matching the distribution of values described by the biometric Law of Ancestral Heredity.[2] The law, which was articulated by Galton and further developed by Pearson, essentially holds that the hereditary contribution of each generation diminishes geometrically by half. For exclusive inheritance, that is the full inheritance of a trait from one parent or another, the law assumes that any offspring had a 50% chance of sharing the character of either parent, a 25% chance sharing the same character as a grandparent, and so on (Pearson, *Grammar*, 1900 489–90). For recessive traits, however, Yule believed the Law of Ancestral Heredity did not apply: "Mendel's laws, so far from being in any way inconsistent with the [biometric] Law of Ancestral Heredity, lead then directly to a special case of that law, for the *dominant* attribute at least. For the *recessive* attribute it does not hold" (Yule 226–27).

In addition to Yule, Pearson entertained the idea that biometrical approaches and Mendelian Genetics were not necessarily exclusive. In the twelfth installment of his "Mathematical Contributions to the Theory of Evolution" series, titled "On a Generalized Theory of Alternative Inheritance with Special Reference to Mendel's Laws" (1904), Pearson used Mendel's ratios as the basis for developing a mathematically supportable description of alternative inheritance. What he concludes is that Mendel's model seems to account reasonably well for the outcomes measured using biometrical methods, though he considered the correspondence of theory to actual measurements insufficient to promote certainty (86).

Though neither Yule nor Pearson refers to Mendelian principles as "causes" of inheritance, their efforts to use them as a legitimate basis for mathematical extrapolation inspired a young Fisher to embrace this conclusion. In a speech comparing Mendelism and biometry given before the second undergraduate meeting of the Cambridge University Eugenics Society in 1911, Fisher states: "It has been shown by Karl Pearson, on whose work the whole science of biometrics has been based, that a number of pairs of Mendelian allelomorphs [or genetic trait pairs] scattered at random in a population would serve as the independent arbitrary causes that biometricians require" (Bennett 56).

This same notion that traits with Mendelian features could serve as causes from which effects could be mathematically extrapolated is echoed in Fisher's first important work, bringing Mendelism and biometry together: "The Correlation between Relatives on the Supposition of Mendelian Inheritance" (1918). As the title suggests, Fisher's general goal in this paper was to examine the biometrical concept of correlation—the mathematically calculated degree of relatedness between relatives—on the assumption that the process of inheritance that linked them was Mendelian. He begins the piece with an implicit reference to the work of Pearson and Yule, but expands the influence of Mendelian conditions in a human population beyond the limits set by the previous authors and shows the value of this expansion: [3]

> Several attempts have already been made to interpret the well-established results of biometry in accordance with the Mendelian scheme of inheritance. It is here attempted to ascertain the biometrical properties of a population of a more general type than has hitherto been examined, inheritance in which follows this scheme. (Fisher, "Correlation" 399) [4]

By expanding Mendelian principles to all cases, Fisher argues, conclusions can be reached about whether and to what extent different genetic and non-genetic causes contribute to the variation of heritable characteristics. This conclusion has important ideological ramifications for both biometricians and Mendelians. On the one hand, it supports the importance of statistical, biometrical methods in testing hypotheses about the influence of various causes on inheritance. On the other, it recognizes the importance of assuming Mendelian causal conditions when investigating inheritance and variation using biometrical methods.

The most important conclusions that Fisher makes are that (1) the environment has little or no effect on the variation of heritable traits,

(2) the dominance of a trait has a constant quantifiable effect, and (3) the other characteristics that heritable traits have been shown to exhibit—specifically epistacy and polymorphism[5]—have no effect on the variability of heritable traits. In order to arrive at these conclusions, Fisher employs a complex mathematical model grounded on several important assumptions. First, he assumes that the total variability of a particular trait in humans is the sum of all the variations of all the Mendelian factors affecting that trait. Second, he assumes that traits are affected by "a large number of Mendelian factors" (400). On the basis of these assumptions, Fisher calculates that the fraternal correlation—the relatedness between brothers—for stature in humans should be 54%. This meant that while, generally, brothers had above a 50% chance of being similar in stature from one another. there was a 46% chance that they would vary.

Perhaps the most significant conclusion Fisher makes in his article is that the 46% chance of genetic variability in brothers was not due to environmental factors as biometricians had suggested:

> For stature the coefficient of correlation between brothers is about .54, which we may interpret by saying that .54 percent of their variance is accounted for by ancestry alone, and that 46 percent must have some other explanation. It is not sufficient to ascribe this last residue to the effects of the environment. (400)

To argue this point, Fisher needed to explain what accounted for the 46% chance of variability in stature if the environment did not. He concludes that much of the variability which was previously attributed to the environment was instead due to the fact that stature was caused by a large number of Mendelian factors. Because many factors were involved, there could also be many possible combinations of factors that would lower the measure of relatedness, thereby increasing the amount of variability amongst offspring:

> The simplest hypothesis and the one we shall examine, is that such features as stature are determined by a large number of Mendelian factors and that the larger variance among children of the same parents is due to the segregation of those factors in respect to which the parents are heterozygous (400).

After postulating that the variability in stature amongst fraternal groups was a consequence of the combining of a large number of Mendelian factors, Fisher devises an ingenious method for proving that environmental factors played a very limited role in inheritance. It had been noted by the biometricians that the correlation amongst fraternal offspring was always higher than the correlation between fraternal offspring and their parents. Assuming a large number of Mendelian factors, Fisher corroborates the biometrical calculation that the correlation between fraternal offspring and their parents was consistently one half less than the fraternal correlation when a dominant genetic factor was involved in a cross.[6] By assessing dominance as a source for increasing variability (as well as the effect of other factors, such as the selective effects of mating preferences in humans), Fisher concludes that almost all of the 46% of non-heritable variation could be accounted for by hereditary factors, with not more than 5% being due to environmental causes:

> By means of the fraternal correlation it is possible to ascertain the dominance ratio and to distinguish dominance from all non-genetic causes, such as environment, which might lower correlations. . . . On this hypothesis it is possible to calculate the numerical influence not only of dominance, but of the total genetic and non-genetic causes of variability. An examination of the best available figures for human measurements shows that there is little or no indication of non-genetic causes. (433)

As we can see from Fisher's very early efforts to develop his mathematical approach to the study of variation and heredity, Mendelian conditions played a central role as the assumed causal conditions on which he built his mathematical arguments. By extending Mendel's principles to broadly encompass heritable traits in a population, Fisher effectively eliminated causation as a problem for the biometrical approach and expanded the common ground between biometry and Mendelian genetics.

Eunoia: Ethos and the Status of Knowledge in Fisher's Mathematical Program

Although Fisher avoided the problem of causation, he faced a second set of obstacles that were more difficult to resolve. First, he had to contend with the fact that most mathematical argument was inaccessible to the majority of biological researchers. Second, he had to deal with the perception amongst biologists that mathematical investigations of variation, evolution, and heredity were simply abstract musings about organisms that could tell them nothing valuable about their actual characteristics.

To understand and explore these two challenges to Fisher's work, it is useful to turn to the rhetorical concept of *ethos* not only as a guide to the conceptual basis of these problems, but also as a heuristic for examining Fisher's efforts to solve them. While chapter five's examination of Galton's ethos offers a relatively broad conception of the appeal as credibility and social status, a more nuanced understanding of its facets is required for the analysis of Fisher's work. In classical and modern rhetorical theory, ethos has three dimensions: *phronēsis* (good sense), *arête* (goodness, excellence, or virtue of any kind), and *eunoia* (good will) (Aristotle, *Rhetoric* II, i1378a 8). Of these three, goodwill plays a central role in the Fisher's struggle to find acceptance for his mathematical program for studying variation, evolution, and heredity.

Eunoia and Benivolentia: Establishing Goodwill with the Audience

What constitutes goodwill towards an audience has had different interpretations in classical Greek and Roman rhetorical traditions. In the examination of Fisher's case, both interpretations are relevant. In the ancient Greek tradition represented in Aristotle's *Rhetoric*, the quality of goodwill is directly linked to the qualities of friendship. Aristotle explains that an audience judges the degree of goodwill in a speaker on the basis of whether or not they believed he or she is being a friend to them. This belief is inspired if the audience feels that the speaker wishes the best for them for their own sake rather than for the sake for some benefit on the part of the speaker (II, iv 1381a 9). It is also inspired when the audience believes that they and the speaker consider the same things good and evil, or that "they are like ourselves in character and occupation" (II, iv 1381b 15).

Aristotle's concept of *eunoia* can be considered the classical equivalent to Burke's more modern concept of *identification*. For Burke, *identification* is the perception in an audience that they share some common interest with a particular speaker or writer. This feeling exists prior to a discourse or can be constructed by a speaker or writer through persuasive tactics in the discourse: "A is not identical with his colleague, B. But insofar as their interests are joined, A is *identified* with B. Or he may *identify himself* with B even when their interests are not joined, if he assumes they are, or is persuaded to believe so" (20).

Whereas Aristotle and Burke understand goodwill as identification between the speaker and the audience, Roman rhetorical theorists discuss it in terms of a speaker's ability to condition his audience to be predisposed to accept his arguments. In his discussion of exordia—openings of speeches whose goal is to predispose the audience towards the speaker and his case—in *De Inventione*, for example, Cicero explains that a beneficially conditioned audience has three characteristics: they are well-disposed, receptive, and attentive to what the speaker is saying (I, xv 20). In order to condition the audience to favorably respond to an orator's case, Cicero suggests that the orator can appeal to their goodwill, or *benivolentia*. Instead of adopting the qualities of a friend, however, he recommends that goodwill can be created if the orator can make himself or his case seem likable while at the same time making the case of his opponent seem disreputable or despicable:

> Good-will is to be had from four quarters: from our own person, from the person of the opponents, from the persons of the jury, and from the case itself. We shall win good-will from our own person if we refer to our own acts and services without arrogance. . . . Good-will may come from the circumstances themselves if we praise and exalt our own case, and deprecate our opponents with contemptuous allusions. (I, xvi, 22)

In addition to relying on the construction of a favorable character for the speaker and the case, Cicero also suggests that the beneficial conditioning of the audience depends on *how* the orator presents the case. He explains that audiences are more disposed to a case if the orator can present it in a manner that makes them attentive and receptive to his arguments. To make the audience attentive, Cicero counsels the orator to "show that the matters which we are about to discuss are important, novel, or incredible, or that they concern all humanity or those in the audience" (I, xvi, 23). To make them receptive, the orator must "explain the essence of the case briefly and in plain language" (I, xvi 23).

In contemporary theories of rhetoric, composition, and communication, Cicero's notion of *benivolentia* as a consequence of *how* arguments are presented emerges in discussions of style and in the creation of accommodated texts. In Deborah Hawhee and Cheryl Crowley's textbook, *Ancient Rhetorics for Contemporary Students,* for example, the authors connect Cicero's discussion of goodwill with composition, suggesting that, like professional writers, students should make their writing both interesting and accessible for their readers (180).

The importance of generating interest and understanding in complex communication is also addressed in Jeanne Fahnestock's "Accommodating Science: The Rhetorical Life of Scientific Facts." Like Hawhee and Crowley, Fahnestock connects accommodation with the ideas of Cicero, concluding that in order to persuade the audience of the importance of scientific achievements, scientific accommodators must "praise or excoriate [some scientific achievement] . . . by attaching it to a category that has a recognized value for an audience," or claim "that something has value because it leads to future benefit" (334).[7] For Fahnestock, scientific accommodators regularly follow Cicero's rules of thumb for audience conditioning by showing that their scientific subjects are "important, novel, or incredible" and by translating scientific ideas into accessible language for their readers.

Because Fisher's work faced a rhetorical situation in which the majority of the audience was both unsure about the benefits of biometry and untrained in its mathematical techniques, both *identification* and *accommodation* presented challenges to argument and persuasion that had to be met. By understanding the characteristics of *eunoia* and *benivolentia*, it is possible to better comprehend some of the critiques leveled against Fisher's early work and the strategies he developed in his later work to gain acceptance for his mathematical approaches to variation, evolution, and heredity.

The Virtue of Practical Knowledge

In addition to meeting the challenges of comprehension and identification, Fisher also had to contend with the basic, pragmatic requirements in early twentieth-century science that knowledge deals with the sensible realities of natural phenomena and, ideally, makes some material contribution to solving the problems of the age. The influence of the former requirement on scientific reasoning is evidenced in the critiques of Pearl and Shull, who challenged the epistemological robustness of the biometrical program on the grounds that it was unbounded by and unconcerned with the physical and ecological realities of actual organisms, and therefore could not be considered practical knowledge by the biologist.

Pearl makes this point bluntly in the *Modes of Research in Genetics* when he writes: "The charge is made that biometrical methods . . . intentionally disregard the detailed study of the individual, and therefore can lead directly to experimental indeterminism as a mode of biological thought" (47). Because Fisher relied on biometrical methods to make his arguments, and supported the study of biological phenomena using theoretical mathematical extrapolation, his work was open to the charge of being simply theoretical with no practical value for understanding nature as it actually was.

A Failure of Ethos: "The Correlation between Relatives"

Fisher's first professional foray into mathematical biology, "The Correlation between Relatives on the Supposition of Mendelian Inheritance," has been considered by many historians and practitioners of population genetics to be Fisher's trailblazing effort to forge a new program of mathematical statistics (Lewontin 58; Provine 144–47). However, a close examination of the initial reception of the text, guided by the ethical dimensions of goodwill and good sense, suggests that, at the time of publication, it was not regarded, for legitimate reasons, as an intellectual tour de force by either biometricians or geneticists.

Although "The Correlation between Relatives" appeared in the *Transactions of the Royal Society of Edinburgh*, the journal had not been Fisher's original choice for publication. The text, originally completed in 1916, was first sent to the *Transactions of the Royal Society*, where its contents were reviewed by two of the leading figures in genetics and biometrics at the time: R.C. Punnett and Karl Pearson. Because these reviewers represented the two intellectual enterprises that Fisher was trying to join with his work, their reviews of his paper and reasons for rejecting it offer critical insight into the conceptual and practical barriers that Fisher faced in trying to make the case for his mathematical program for the study of variation, evolution, and heredity.

Perhaps the most obvious issue that emerged in the reviews of his first article was the problem of accommodation. The geneticist R.C. Punnett—a protégé of Bateson's—wrote in his review that he was at a loss to understand either Fisher's mathematics or his biology. About the former, he writes "I have had another go at this paper but frankly I do not follow it owing to my ignorance of mathematics" (qtd. in Norton and Pearson 154). Although this response could be rightly interpreted as a typical reaction for geneticists who had generally very little training in complex mathematics, Punnett had some interest in and experience with mathematical biology being the biologist who sought out Hardy's mathematical advice for answering the question of dominant traits in eye color, which eventually lead to the development of the mathematical Hardy-Weinberg principle.

In addition, Punnett's confusion is supported by later commentaries on Fisher's work by experts in statistical biology, who, looking back on the paper, also remarked on the turgidity of its statistical arguments. A reviewer of a 1967 reprint of Fisher's paper writes:

> The commentary [on the original paper] shows how many of the results Fisher uses can be [mathematically] derived—one hesitates to say were derived when referring to Fisher's writing. . . . He was a very economical writer frequently contenting himself with no more than sketching even important parts of his argument. (Mather 372)

In addition to criticizing the elusive meaning of Fisher's mathematical argument, Punnett also complained in his review that Fisher was not clear about his biological explanations:

> I don't understand what he means by the expression 'that there is in dominance a certain latency' & the two sentences following. They are evidently important to his conclusion & though they are no doubt perfectly clear to him I feel that he ought to illustrate what he is driving at by a simple example. In short for the sake of the biologist he must translate this part into the language which the biologist habitually uses. (qtd. in Norton and Pearson 155)

Besides failing to establish goodwill with his readers by making his mathematical and biological arguments accessible, Fisher is also unsuccessful in convincing them that his investigations had the virtue of practical knowledge. The second reviewer, Karl Pearson, for example, was skeptical of the general value of the paper because Fisher had not collected any empirical data to support his claims. Without knowledge of particular instances, Pearson argued, Fisher's paper was restricted to mathematical hypothesizing that held no great interest for biometricians who were trying to measure and understand real populations: "I do not think in the present state of affairs the paper is wide enough to be of much interest from the biometric standpoint for its hypotheses need some observational basis" (qtd. in Norton and Pearson 154).

Whereas Pearson's critique of the practicality of Fisher's work focuses on the paper's lack of quantitative observational data, Punnett's assessment challenges this virtue on the ground that its assumptions preclude the possibility of experimental verification, thereby putting it outside the range of practical knowledge. A particular point of contention for both Punnett and Pearson was Fisher's assumption that stature was made up of a large number of Mendelian factors. Geneticists had already considered characters to which two or three different factors contributed and had done experiments to calculate the effects of those factors using Mendelian ratios. These efforts, however, were extremely laborious. By assuming a large number of Mendelian factors, Fisher's work proposed a hypothesis beyond the reach of experiment and, therefore, beyond the capacity to produce the type of practical knowledge of organisms valuable to geneticists. Recognizing this problem, Punnett writes:

> I should doubt therefore that assumption (3) [that a given character is a manifestation of a large number of Mendelian factors] could ever be tested by experiment. It may be true and it may not be and at present I don't see how one can say any more. . . .Whatever its value from the standpoint of statistics & population, I do not feel that this kind of work affects biologists much at present. It is too much of the order of problem that deals with weightless elephants on frictionless surfaces, where at the same time we are largely ignorant of the other properties of the said elephants and surfaces. (qtd. in Norton and Pearson 155)

An examination of the reception of Fisher's earliest attempts to bridge the gap between genetics and biometrics reveals that his efforts were rejected by both sides. On the one hand, the geneticist reviewers complained that Fisher showed insufficient goodwill towards his geneticist audience by not making his biological and mathematical ideas accessible. On the other, both the genetic and biometrical reviewers opined that the work lacked the virtue of practicality.

Following its rejection from the *Transactions of the Royal Society*, "The Correlation between Relatives" remained unpublished for two years. That it eventually appeared in the *Transactions of the Royal Society of Edinburgh* is more a consequence of Leonard Darwin's personal enthusiasm for Fisher's work than of a general sentiment amongst geneticists or biometricians that it should be published. In 1918 he convinced the Eugenics Education Society to sponsor the paper. Their search for a willing publisher led them to the *Transactions of the Royal Society of Edinburgh*, who agreed to publish the paper but only on the condition that the interested parties raise £43 to cover the costs of printing (Bennett 68–69).

With financial contributions from Fisher, Leonard Darwin, and the Eugenics Education Society, "The Correlation between Relatives" finally appeared in the fall of 1918; however, it initially made little if any impact in biometrics or genetics until decades after its initial publication. A search of the citations of the article in a range of scientific journals in JSTOR reveals that, between 1918 and 1950, there were only three citations of the article in work of a statistical or biometrical nature that did not appear in papers by Fisher or Leonard Darwin (Haldane, "Some Statistical Problems;" Jones; and Kelley).[8] If Fisher's and Leonard Darwin's works are included, the citation rate increases to five.[9] It is not until the decade between 1953 and 1963 that "The Correlation between Relatives" specifically, and Fisher's work generally, begins to be discussed in the manner in which it is considered today, as a major step in the synthesis between genetics and biometry.

Blurring Borders: The Importance of Ethos in Building Fisher's Mathematical Program of Biological Investigation

Given the abstractness of Fisher's earliest work, a few philosophers and historians have puzzled over how "Correlation between Relatives" could have provided the basis for the modern science of population genetics which deals increasingly with the qualities and relationships of real populations (Plutynski; Provine). In her analysis of Fisher's 1918 paper, for example, philosopher Anya Plutynski argues that Fisher was successful despite the abstract quality of his work because he was able to provide "proof of possibility" that there was a connection between Mendelian genetics and Darwinian natural selection (1207–08). This proof was sufficient to encourage statisticians and biological researchers to initiate a mutually supportive program of investigations despite the theory's lack of details or evidence from the natural world.

The initial reception of Fisher's paper and its consequent citation record suggest, however, that such an enthusiasm over the proof of possibility was not present. If researchers in both genetics and biometry were not compelled by the possibilities of his work, how were they eventually persuaded to see his mathematical approach as a reasonable means for a unified explanation of variation, evolution, and heredity? The answer to this question has important implications for understanding not only Fisher's particular efforts to ameliorate his methods and ideas to his primary audiences, but also the general process in science by which a common identity is shared by initially separate scientific enterprises. This section suggests that ethical strategies and argumentation play an important role in both.

Lessons Learned: Rothamsted Station and "On the Dominance Ratio"

Fisher's struggle to get the "The Correlation between Relatives" published offered a valuable lesson for the novice arguer about the importance of ethos to the development of his mathematical program of investigation. However, it was not his only source of instruction on this issue. There were a variety of conditions and experiences that led not only to a growing rhetorical consciousness, but also to an increase in his rhetorical skills. By exploring Fisher's professional endeavors in the years following his first publication, and examining the text of his second major publication, "On the Dominance Ratio," it is possible to get sense of the conditions which affected his rhetorical awakening and how Fisher drew on these experiences to establish a credible ethos for his work.

In 1918, weeks after publishing "The Correlation between Relatives," Fisher announced his intentions to leave his job at Bradfield College, a secondary school, in hopes of finding a position either in higher education or research. After almost a year of unsuccessful searching, he was offered two positions: one from Rothamsted Experimental Station, and one from Karl Pearson to do statistical work at the Galton laboratory. Although Pearson's offer would have given Fisher the chance to pursue his interests in statistics at the highest professional level, he turned it down because of professional conflicts that had developed between him and Pearson.[10]

Fisher's choice of the job at Rothamsted put his career path on a whole new trajectory. The experimental station was devoted to the pragmatic pursuit of the scientific study of agriculture. In accepting the position, Fisher chose to turn his talents toward assessing the practical utility of statistics to agriculture rather than to the expansion of the theoretical boundaries of statistical mathematics. His situatedness in this kind of intellectual environment had important consequences for his rhetorical development. On the one hand, it placed him in a position where establishing the practical value of statistics was paramount. According to Fisher's daughter and biographer Joanne Box, the survival of Fisher's position at Rothamsted beyond the initial six-month-to-one-year appointment, "depended on whether he could prove the usefulness of his particular contribution to agriculture: of statistics to practical research men" (61).

On the other, it forced Fisher to carefully consider the challenges that complex statistical methods posed to non-specialist audiences. Part of this awakening was the result of the working conditions at his new assignment. Unlike Cambridge or the Galton Biometric Laboratory, the experimental station could supply him only with a simple calculating machine and a single assistant. Without access to more sophisticated computing power or assistants, Fisher had to develop less sophisticated methods for doing analysis. Another part of his awakening was the result of his recognition that his audience, agricultural specialists, found themselves in similar or worse conditions, which seriously limited their capacity to carry out complex computations on a routine basis. Box explains:

> His own experience at Rothamsted gave an edge to his ingenuity in avoiding unnecessarily laborious or complicated arithmetical methods. . . . He was constantly made aware of the need, in practice, for a means of checking arithmetical results, especially when the same sorts of calculations were made frequently in routine work. (109)

Working outside the specialist confines of academic genetics and biometrics supplied Fisher with exigencies and experiences that gave him a more robust insight to the importance of accommodation and practicality to the success of his arguments. In a letter to the geneticist Cyril Darlington on January 6, 1936, Fisher recognized the importance of his engagement with non-specialists audiences to his development as a scientist-arguer:

> I am convinced that publication in non-specialist journals has been very much to my personal advantage, both in forcing me to think out problems from the point of view likely to need their solutions, and in bringing my methods to the notice of a far wider group of workers likely to use them. (qtd. in Bennett 190)

By 1922 there are signs that Fisher's experiences at Rothamstead and the difficulties he faced in getting "The Correlation between Relatives" published were beginning to influence the argument strategies in his work. Changes in his tactics can be recognized in his next important effort, "On the Dominance Ratio," in which he synthesized Mendelian heredity, biometry, and Darwinian natural selection. The paper, which appeared in the *Proceedings of the Royal Society of Edinburgh*, includes features that respond, in some cases directly, to the ethical shortcomings of his 1918 paper.

In the first three pages, Fisher opens the work with a discussion whose primary goal is to make the case that the foundational assumptions of his paper, "The Correlation between Relatives," have the virtue of practicality. In particular, he addresses the criticism leveled by both Punnett and Pearson that there is no empirical evidence to support his contention that a large number of Mendelian factors make up any given physical trait in an organism. Fisher refutes this critique by citing evidence from genetic researchers who corroborate his assumption:

> At the time when the 1918 paper was written, it was necessary . . . to show that the assumption of multiple, or cumulative factors, afforded a working hypothesis of the inheritance of such apparently continuous varieties of human stature. This view is now far more widely accepted. . . . In some fortunate circumstances, as in *Drosophila*, it has been possible to isolate and identify the more important of these factors by experimental breeding on the Mendelian method. ("On the Dominance Ratio" 322)

Although presenting experimental instances that support his assumption of a large number of Mendelian characteristics might have been sufficient to persuade readers of the practical value of his mathematical model, Fisher was intent on making a much bolder conclusion. He claimed that if mathematical arguments have been found to be practically useful in this case, surely, we can depend on them to be useful in other cases even if there is no way to independently verify their conclusions through experimentation. To make this argument compelling, he adds that granting this quality to mathematical modeling would allow the practical exploration of heredity and variation in important domesticated organisms in which lengthy and intricate experimental studies, like the ones done on *Drosophila*, were impossible:

> More frequently, however, and especially in the case of the economically valuable characters of animals and plants, no such analysis [through breeding experiment] has been achieved. In these cases we can confidently fall back upon statistical methods, and recognize that if a complete analysis is unattainable, it is also unnecessary to practical progress. (322)

In these lines, Fisher offers readers hope that by trusting in the ontological validity of mathematical models they can overcome the enormous obstacles of having to experiment on complex and economically important organisms. In doing so, Fisher hopes to lure biological researchers beyond the particular case of his previous paper towards the conviction that the whole enterprise of mathematically modeling variation, evolution, and heredity is endowed with the virtue of practicality. With this proposition, he opens his new venture to combine biometry, Mendelism, and Darwinism.

In addition to arguing that his previous and current work has the ethical virtue of good sense, Fisher also appeals in "On the Dominance Ratio" to the ethical dimension of good will by accommodating the mathematical information in the text for his audience of conventional biological researchers. Though the arguments still include mathematical components that would have challenged most geneticists, Fisher attempts to make his discussion more accessible by scaling back his use of mathematical arguments and by visually presenting his main points.

Perhaps the most profound accommodation that Fisher makes in the 1922 paper is the extent to which he limits his mathematical reasoning. A count of the number of lines of mathematical calculations in Fisher's 1918 correlations paper nets an approximate total of 246 lines of calculations, which averages to about 7.2 lines for each of the thirty-four pages of text. In contrast, "On the Dominance Ratio" includes approximately 96 lines of calculations in twenty pages of text, for an average of 4.8 lines per page. On the whole, therefore, Fisher reduces the number of lines of mathematical calculation by one third. Although comparing Fisher's 1918 paper with his 1922 efforts is perilous because the two papers do not examine in all cases the same subject in the same way, the substantial difference between the use of mathematics in the two papers and the relative similarity in their topics suggests that this discrepancy is a reasonable indication of a change in Fisher's approach.

In addition to limiting the amount of mathematical reasoning, Fisher also appears to make efforts to accommodate geneticists by presenting his most important mathematical conclusions in an accessible graphical form. One of the central issues that Fisher addresses in this paper is whether and to what degree different variables such as population size, the presence or absence of natural selection, and the presence or absence of genetic dominance affects the degree of variability within a population. In order to ascertain these effects, he adopts three different models: one in which the variability is affected only by chance, one in which it is affected only by natural selection, and one in which it is affected by both dominance and natural selection. Mathematically, Fisher works out the results for each set of circumstances and concludes that the model that includes both dominance and natural selection is most representative of the actual condition of variability in nature (340).

Although Fisher could have let the mathematical comparisons of these models speak for themselves, he takes extra measures to translate the results into graphics that illustrate his conclusions. For the each of the models, he provides a curve which shows the distribution of variability in a random population after a generation of breeding under the conditions specified by the model. In the first of two sets of visuals, for example, Fisher compares two curves, which reveal for the reader the effect of random extinction of genes without replacement through mutation. The first curve is a prototypical, symmetrical bell-shaped curve showing the distribution of the variability of a population in which gene types that are randomly eliminated are replaced by the introduction of new mutations (Figure 2). In the next graphic on the same page, he presents what happens to variability when gene

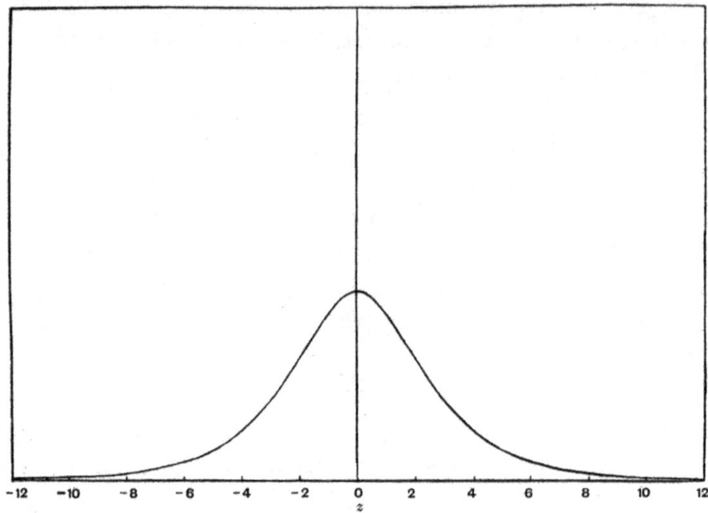

Distribution of the logarithmic frequency ratio $\left(z=\log\frac{p}{q}\right)$ of the allelomorphs of a dimorphic factor.

FIG. 1.—$df=\frac{1}{2\pi}\operatorname{sech}\frac{1}{2}z\,dz$;

represents the distribution when, in the absence of selection, fortuitous extinction is counterbalanced by mutation. Dominance Ratio = ·2308.

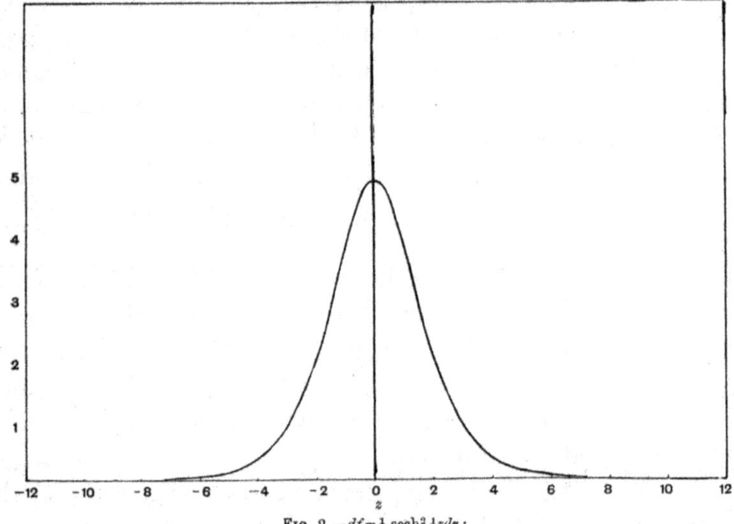

FIG. 2.—$df=\frac{1}{4}\operatorname{sech}^2\frac{1}{2}z\,dz$;

represents the distribution when, in the absence of selection and mutation, the variance is steadily decaying owing to fortuitous extinction of genes. Dominance Ratio = ·2500. This is the condition emphasised by Hagedoorn.

Figure. 2. Curves Comparing the Rand om Extinction of Genes with and without Replacement. Reprinted from R.A. Fisher, "On the Dominance Ratio," *Proceedings of the Royal Society of Edinburgh* 42, p. 329. © 1922. Used by Permission of the Publisher, The Royal Society of Edinburgh.

types eliminated by random extinction are not replaced. From the visuals it is easy to see that the amount of variation in a population shrinks considerably. The central hump of the curve, representing the norm, thins and rises while the lines at the flared edges, representing variation from the norm, shrink and lose altitude as the population becomes significantly less variable.

As a supplement to the visuals, Fisher supplies readers with descriptive captions that help them connect the graphic with the biological condition they are intended to describe. In the second set of graphs, for example, he adds, along with the formula described by the graph, the biological condition: "genetic selection counterbalanced by mutation" (333) (Figure 3). This descriptor allows the reader to distinguish the conditions responsible for the mesa-shaped distribution on the top graph from the right of center, the skewed distribution represented in the bottom graph, which is caused by "genotypic selection, with complete dominance, counterbalanced by mutation" (333) (Figure 3). Finally, Fisher includes in his caption for the bottom graph a reiteration of the paper's general conclusion, which is that this particular set of conditions is the one that represents the true state of nature. He writes "This is the probable condition of natural species, including man. Note the accumulation of rare recessives" (333).

Although graphs also appear in Fisher's 1918 paper, there is a marked difference in the type of visuals and the manner in which they are presented to the audience. These differences suggest that Fisher had more seriously considered the utility and accessibility of his visuals to geneticists in crafting the 1922 article. One of the most obvious differences between visuals in the two papers is the way the captions describe the graphic representations. Whereas his 1922 paper employed natural language descriptors to explain the biological relationships or conditions presented by the visuals, his 1918 paper presented information about the graphics in terms of abstract mathematical variables, which stood for some factor effecting variability, such as dominance, homogamy, etc. Figure 4, for example, describes the distribution of the dominance ratio in a population with Mendelian inheritance (165).

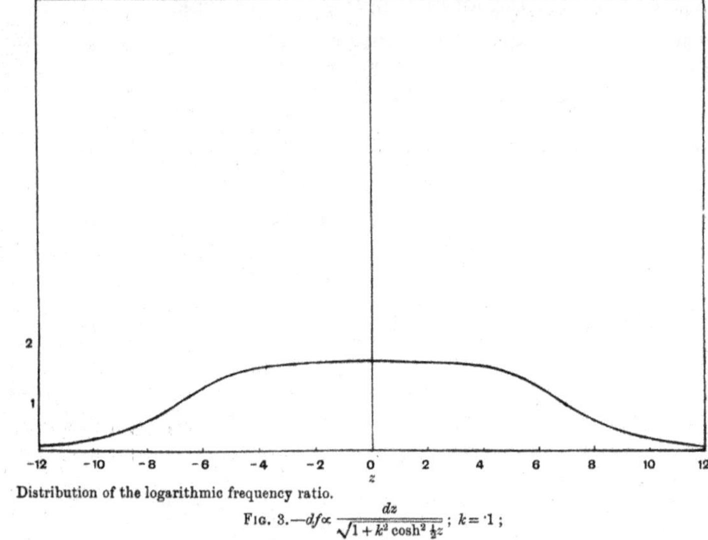

Distribution of the logarithmic frequency ratio.

FIG. 3.—$df \propto \frac{dz}{\sqrt{1+k^2 \cosh^2 \frac{1}{2}z}}$; $k = \cdot 1$;

genetic selection counterbalanced by mutation. Dominance Ratio, ·2000.

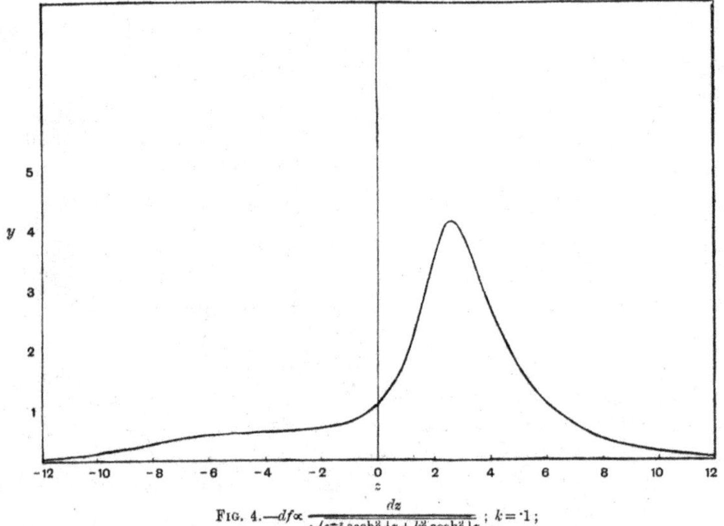

FIG. 4.—$df \propto \frac{dz}{\sqrt{e^{-z} \operatorname{sech}^2 \frac{1}{2}z + k^2 \cosh^2 \frac{1}{2}z}}$; $k = \cdot 1$;

genotypic selection, with complete dominance, counterbalanced by mutation. Dominance Ratio, ·3333. This is the probable condition of natural species, including man. Note the accumulation of rare recessives.

Figure 3. Curves Comparing Genetic Selection Counterbalanced by Mutation and Genetic Selection with Complete Dominance Counterbalanced by Mutation. Reprinted from R.A. Fisher, "On the Dominance Ratio," *Proceedings of the Royal Society of Edinburgh* 42, p. 333. © 1922. Used by Permission of the Publisher, The Royal Society of Edinburgh.

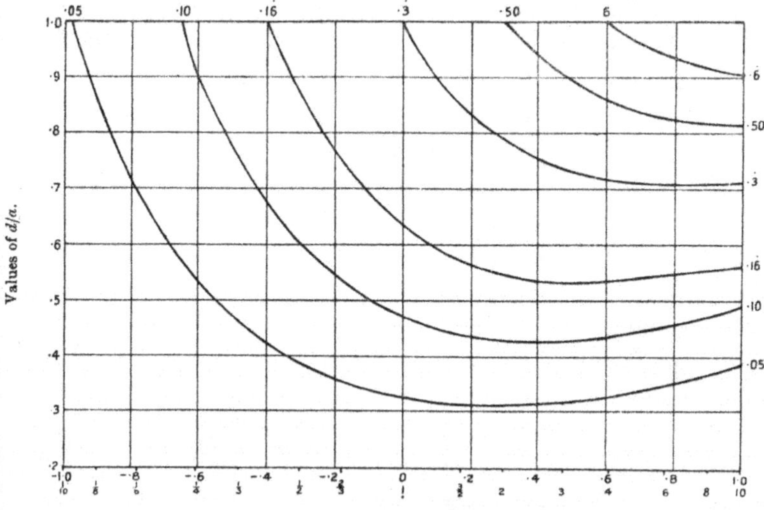

FIG. 1.—Values of $\log_{10}(p/q)$ (upper figures) and of p/q (lower figures).

Figure 4. The Dominance Ratio in a Mendelian Population. Reprinted from R.A. Fisher, "The Correlation between Relatives," *Transactions of the Royal Society of Edinburgh* 52, p. 430. © 1918. Used by Permission of the Publisher, The Royal Society of Edinburgh.

Unlike the graphics in "On the Dominance Ratio," there is no natural language description in the caption of the graph which explains the general relationship it shows. Instead, the reader, glancing only at the chart is forced to decipher this relationship from the fact that the value of the curves in the graph describe a relationship between *p*/*q* on the *x* axis of the graph, and *d*/*a* on the *y* axis. If the reader comes to the chart from the text, the clarity of the visual is only slightly improved. Unless he is able to wade through Fisher's abstruse mathematical argument, the reader can only glean from Fisher's discussion that the graph shows that the value for the parental correlation ratio can be above or below .3, which is a challenge to the Law of Ancestral Heredity (428). The important ramifications of this difference for understanding heredity, however, remain opaque. The other three graphics are similarly presented, making them a challenge to all but a very few, highly trained mathematical readers. By using natural language descriptors to highlight the important biological relationships in the graphs in his later work, Fisher made a substantive improvement in the accessibility of his arguments to these audiences.

A brief comparison of Fisher's early efforts to synthesize Mendelism and biometry in "The Correlation between Relatives," and biometry, Mendelism, and Darwinism in "On the Dominance Ratio," reveals that he made important changes to the way he presented his arguments and his data. To establish the virtue of practicality, Fisher revisited the theoretical assumptions supporting his mathematical model in the 1918 paper and vindicated them by identifying experimental research that had consequently confirmed his assumptions. On the basis of this vindication, Fisher also made a bolder claim to practicality by arguing that his methods had real practical applicability in cases where experimental verification would either be impossible or cumbersome. Finally, he makes efforts to garner the goodwill of the audience by reducing the amount of mathematics and by framing his discussion from a biological perspective, providing graphical and natural language support to make his arguments and conclusions more accessible.

Reaching Across Disciplinary Divides: The Genetical Theory of Natural Selection

By the late 1920s, "On the Dominance Ratio" had made sufficient headway in bringing together the ideas of Mendel, Darwin, and the biometricians that Fisher was able to strike a deal with Oxford's Clarendon Press for a book on the subject. While his articles had generated interest in the development of his biometric method amongst a small group of specialists, the prospects of a book gave him the opportunity to appeal to a much broader audience. In this context, establishing the ethos for his work became a paramount concern. If he could convince a wider group of geneticists, biometricians, and other biological researchers that mathematical approaches to biology could produce useful knowledge, and if he could predispose them to make efforts to understand or participate in a new type of mathematical biology, he might be able to establish a strong program of biomathematical research.

Preface: The Practical Benefits of a Biological/Mathematical Enterprise

Because of his previous experiences with conventional biologists, Fisher knew that in order to get hearing for his ideas in the book, he needed to establish that his mathematical study of variation, evolution, and heredity had the virtue of practicality. To respond to this ethical challenge, Fisher presents his readers with an exordium in the preface of *The Genetical Theory of Natural Selection.* In the exordium he endeavors to characterize his mathematical enterprise as both profitable and generative of practical knowledge, using a combination of historical exempla and philosophical argument.

In the opening of the preface, Fisher offers the reader a historical account of the triumphs and failures of scientific research that has culminated in the current state of research into variation, evolution, and heredity. This teleological narrative illustrates the importance of mathematical imagination in scientific research and reassures readers of the practical virtue of his proposed program of biomathematical research.

The narrative begins with Darwin, whose most important virtue, Fisher argues, was his capacity for explaining evolution on the basis of natural selection, a phenomena which could be empirically and quantitatively assessed:

> For advocates of natural selection have not failed to point out, what was evidently the chief attraction of the theory to Darwin and Wallace, that it proposes to give an account of the means of modification in the organic world by reference only to 'known,' independently demonstrable, causes. The alternative theories of modification rely, avowedly, on hypothetical properties of living matter which are inferred from the facts of evolution themselves. (vii)

By identifying *The Genetical Theory of Natural Selection* as a work whose origins have their root in Darwin, and by characterizing Darwin's accomplishment as his capacity to explain variation on the basis of "'known,' independently demonstratable, causes," Fisher attempts to establish empiricism as a primary quality for his own enterprise. Just as Darwin embraced only independently verifiable causes, so Fisher also supports mathematical extrapolation that can be verified by experimental and observational evidence.

Following Darwin, the next great historical figure to contribute to the understanding of natural selection was Gregor Mendel, who supplied researchers with the theory of heredity that Darwin lacked. Fisher's description of Mendel's work and its reception provides the reader with an exemplum or anecdote illustrating the practical nature of mathematical reasoning and the importance of developing an interdisciplinary program of research. Of Mendel and his success, Fisher writes: "It deserves notice that the first decisive experiments which opened out in biology this field of exact study [the field of genetics], were due to a young mathematician, Gregor Mendel, whose statistical interests extended to the physical and biological sciences" (viii).

In this sentence, Fisher emphasizes the centrality of Mendel's mathematical training and methods by characterizing Mendel as a "young mathematician" with "statistical interests" rather than as a natural philosopher or botanical/horticultural researcher. This characterization emphasizes the importance of mathematical reasoning in the creation of experimental genetics. This recognition of the original intertwinement of these two enterprises not only suggests a natural compatibility between them, but also implicates mathematics as a fundamental component of a science identified with the actual and the practical.

Whereas Fisher uses Mendel's mathematical predispositions to suggest a functional compatibility between mathematics and genetics, he turns to the tragic disregard of and dispute over Mendel's work as a lesson about the perils of refusing to recognize the affinities and contributions of biology and mathematics to the study of genetics. Recounting how Mendel's genius was dismissed by his contemporaries, Fisher writes: "It is well known that his experiments were ignored . . . and it is presumed that they were never brought under the notice of any man whose training qualified him to appreciate their importance" (vii). Here Fisher attributes the neglect of the most important work in genetics to the failure of the biological community to educate themselves in mathematics or to invite into their scientific enterprise individuals trained in mathematics. Although, as an ethical strategy, accusing the portion of your audience from whom you most need to win acceptance of incompetence and myopic exclusiveness is a perilous undertaking, Fisher neutralizes the sting of his accusation by placing an even greater share of the guilt on mathematical biologists who had the training to understand Mendel's work, but still resisted it:

> It is no less remarkable that when, in 1900, the genetic facts had been rediscovered . . . and the importance of Mendel's work was at last recognized, the principle opposition should have been encountered from the small group of mathematical statisticians [the biometricians] then engaged in the study of heredity. (viii)

By blaming both mathematical and conventional biologists for the rejection of Mendel's work, Fisher does more than chastise each group for their particular nearsightedness. Instead, he places blame on both groups for their collective refusal to recognize the importance of working with one another and illustrates how such uncooperativeness can lead to unnecessary delays and obstacles in the development of science. He argues that if this attitude is left uncorrected, more such scientific tragedies will occur.

Once Fisher has sufficiently prepared his readers with historical examples to recognize the empirical roots of his mathematical program of investigation and the follies of not working cooperatively together, he turns his attention to discussing exactly what sorts of contributions each side can expect from the other in the new scientific enterprise he is promoting. Toward this end, he presents readers with a brief, philosophical discussion, contrasting the training of the biological and mathematical intellect. On one level, this comparison helps readers understand the contributions that each type of training makes in the production of natural knowledge. However, on another level, the discussion challenges critiques that a mathematical program for the study of variation, evolution, and heredity could not make substantial contributions to the advancement of practical knowledge.

In the preface of *The Genetical Theory of Natural Selection*, Fisher challenges his audiences' fundamental assumption that reliable, practical knowledge of the natural world can only be attained through experience with natural phenomena. He opens his case by making his audience of non-mathematical biologists aware that this prejudice exists amongst them: "Most biologists probably feel that this advantage [of the training of the imagination to develop original thoughts about nature] is all on their side. They are introduced early to the immense variety of living things . . . at the same time when the mathematician seems to be dealing with only the barest abstractions" (vii).

Once he has placed this prejudice before the eyes of the reader, he makes an argument from authority to challenge it. His authoritative source is the astronomer Arthur Stanley Eddington, who was an illustrious man of science and well-known in the public as a writer of a number of popular books about science and astronomy. Quoting from a work Eddington had recently published, *The Nature of the Physical World*, Fisher presents the reader with an alternative perspective from which to consider the character of mathematical knowledge: "'We need scarcely add that the contemplation in natural science of a wider domain than the actual leads to a far better understanding of the actual'" (vii). The argument epitomized by these lines is that mathematics leads to a better understanding of the real by creatively expanding our thinking beyond what we actually know to be true from experience.

Once he has made an effort to construct a positive ethos for mathematics by arguing for its practical virtues, Fisher asks his audience to reconsider the misconception that mathematics is limited to the imaginary and theoretical. By expanding their characterization, he argues, both biologists and mathematicians could embark on a program of mutual and, if we consider the Mendelian example, beneficial cross-border collaboration.

> The most serious difficulty to intellectual co-operation would seem to be removed if it were clearly and universally recognized that the essential difference [between mathematics and biology] lies, not in intellectual methods, and still less in intellectual ability, but in an enormous and specialized extension of the imaginative faculty, which each has experience in relation to the needs of his special subject. (ix)

By calling on his readers to recognize that the difference between mathematical knowledge and biological knowledge is an issue of scope of attention rather than quality, Fisher hopes to reassure them that working with mathematicians is both philosophically justifiable and mutually profitable.

Goodwill and Identification in *The Genetical Theory of Natural Selection*

Whereas the preface of *The Genetical Theory of Natural Selection* was dedicated to predisposing Fisher's readers to the idea of a joint enterprise between mathematical and conventional biology, the title and contents of the text were specifically designed to attract the audience to participate in this new enterprise. Perhaps the most explicit evidence that Fisher hoped to excite his readers' interests in his program of mathematical biology was his choice of title for the text. In the early correspondences between Fisher and his editor, Kenneth Sisam, the latter inquired about what Fisher intended to title the book. In a letter dated May 14, 1929, Fisher responded:

> I should call the book something like THE GENETICAL THEORY OF NATURAL SELECTION. I cannot easily get the words mathematical or statistical into the title but genetical is essential. My impudence in treating the subject as a branch of mathematics, I must justify in a preface. (qtd. in Bennett 19)

In his recognition that adding the words "mathematical" or "statistical" could not be done easily, Fisher reveals that he understands their inclusion might alienate a broader audience of genetic researchers who would take his efforts to study variation and heredity mathematically as "impudent." Further, he divulges his desire to attract such an audience by insisting that the word "genetical" was essential to the title. This term would immediately recommend the contents of the book to geneticists and encourage them to identify with its mission.

In addition to naming the book strategically to attract geneticists, Fisher also reveals that he has thought about how to establish *benivolentia* with his audience by making the contents of the book as simple to access as possible. In response to a suggestion from Sisam that the book needed to be simplified to reach a larger audience, Fisher explains that he had already thought a great deal about how to improve its accessibility and that he was willing to further change the manuscript to accommodate his readers: "Many thanks for your letter of May 28. I shall do my best to improve the presentation in the way you suggest, though of course most of what you say is so probably true that I have worried about it a good deal already" (qtd. in Bennett 20).[11]

Evidence that Fisher made efforts to accommodate a broader class of researchers in hopes of getting them interested in his mathematical approach can be found by examining some of the stylistic features of the book. As in "On the Dominance Ratio," Fisher takes pains not to scare away the portions of his audience who might have some mathematical competence, but would otherwise have trouble following the particulars of his argument without guidance. In the most mathematical portion of the book, chapters two through five, Fisher restricts the use of complex mathematical formulae to approximately 203 lines, which averages out to over 112 pages to roughly 1.8 lines per page. In comparison, this is fewer than the number of lines of complex mathematics in all of "The Correlation between Relatives," and a sixty percent reduction of the average number of lines per page in "On the Dominance Ratio."

In addition to limiting the quantity of mathematics in the text, Fisher importantly provides his readers with more explanatory support using natural language to help them understand the meaning of his mathematical formulae. A comparison between the descriptions of his formula for the chance of the survival of an individual gene in "On the Dominance Ratio," and in *The Genetical Theory of Natural Selection* offers direct evidence of these differences. In his 1922 paper, Fisher presents the formula in the following manner:

2. The Survival of Individual Genes.

If we consider the survival of an individual gene in such an organism as an annual plant, we may suppose that the chance of it appearing in the next generation in 0, 1, 2, 3 individuals to be

$$p_0, p_1, p_2, \ldots$$

where

$$p_0 + p_1 + p_2 + \ldots = 1.$$

If

$$f(x) = p_0 + p_1 x + p_2 x^2 + \ldots$$

then evidently if there were two such genes in the first generation, the chance of occurrence in r individuals, or more strictly, in r homologous loci, in the second generation, will be the coefficient of x^r in

$$(f(x))^2.$$

It follows that the chance of a single gene occurring in r homologous loci, in the third generation, will be coefficient of x^r in

$$f(f(x)).$$

Figure 5. Equations Plus Explanations Describing the Survival of an Individual Gene in "On the Dominance Ratio." Reprinted from R.A. Fisher, "On the Dominance Ratio," *Proceedings of the Royal Society of Edinburgh* 42, p. 325. © 1922. Used by Permission of the Publisher, The Royal Society of Edinburgh.

The exact same explanation is given in *The Genetical Theory of Natural Selection*, but includes a more detailed discussion of what the equation represents and how Fisher arrives at the final equation $f(f(x))$:

The chance of survival of an individual gene

An individual gene carried by an organism which is mature, but has not reproduced, will reappear in the next generation in a certain number 0, 1, 2, 3 etc. of individuals or homologous loci. With bisexual organisms these must of course be separate individuals, but where self-fertilization is possible the same gene may be received by the same individual offspring in each of its two parental gametes, and if such an individual survives to maturity our original gene will thus be doubly represented. In general we shall be concerned with the total number of representations, although it will be convenient to speak as though these were always in different individuals. The probabilities that of the offspring receiving the gene, 0, 1, 2 . . . attain maturity will be denoted by

$$p_0, p_1, p_2, \ldots\ldots,$$

where, since one of these contingencies must happen,

$$p_0+p_1+p_2+\ldots\ldots=1$$

In order to consider the chances in future generations we shall first calculate the appropriate frequencies for the case in which our gene is already represented in r individuals. In order to do this concisely we consider the mathematical function

$$f(x)=p_0+p_1x+p_2x^2+\ldots\ldots$$

This function evidently increases with x from p_0, when $x=0$, to unity when $x=1$. Moreover, if the r individuals reproduce independently, the chance of extinction in one generation will be p_0^r. The chance of representation by only a single gene will be

$$rp_0^{r-1}p_1,$$

and in general the chance of leaving s genes will be the coefficient of x^s in the expansion of

$$(f(x))^r.$$

Now, starting with a single gene, the chance of leaving r in the second generation is p_r, and the chance that these leave s in the third generation will be the coefficient of x^s in

$$p_r(f(x))^r.$$

It follows that the total chance of leaving s in the third generation, irrespective of the number of representatives in the second generation, will be the coefficient of x^s in

$$p_0(f(x))^0+p_1(f(x))^1+p_2(f(x))^2+\ldots\ldots,$$

or, in fact, in

$$f(f(x)).$$

Figure 6. Equations Plus Explanations Describing the Survival of an Individual Gene in T*he Genetical Theory of Natural Selection*." Reprinted from R.A. Fisher, T*he Genetical Theory of Natural Selection*," pp 73–74. © 1930. Used by Permission of the copyright holder, University of Adelaide.

In the explanation from *The Genetical Theory of Natural Selection* (Figure 6), for example, Fisher makes a greater effort to explain the first equation "$f(x) = p0 + p1x$. . . ." In "On the Dominance Ratio," the equation is presented with no description for the audience of what the mathematical line means (Figure 5). This presentation format requires the reader to have rigorous mathematical training or have no hope of understanding what Fisher is trying to accomplish in this step. In *The Genetical Theory of Natural Selection,* however, Fisher offers a brief explanation of what the mathematical line accomplishes: "In order to consider the chances in future generations we shall first calculate the appropriate frequencies for the case in which our gene is already represented in r individuals. In order to do this concisely we consider the mathematical function" (74). By explaining what the mathematics accomplishes, Fisher helps the reader with some mathematical competence follow the arc of his reasoning, regardless of whether the reader can fully follow the mathematics. This additional explanation in *The Genetical Theory of Natural Selection* is evidence of Fisher's desire to establish good will in the audience, to reach across borders to get geneticists interested in his mathematical program of argument.

A close textual analysis of the preface, title, and content of *The Genetical Theory of Natural Selection* reveals Fisher's efforts to reach across disciplinary borders to get geneticists interested in his mathematical program of argument. In the preface of the text, he attempts to predispose his readers to his case by showing them the benefits and costs of a combined program of mathematical-biological science and by arguing for the practical virtue of a mathematical program of investigation. Through choices in the body and title of the text, Fisher strives to make the content of the book interesting and accessible by carefully choosing which aspects of the subject matter to highlight, limiting mathematical argument, and providing natural language descriptions.

Reader Response to *The Genetical Theory of Natural Selection*

The preceding sections of this chapter have highlighted Fisher's efforts to establish a credible ethos for his mathematical investigations of variation, evolution, and heredity. However, the question still remains, "Did these strategies succeed in producing the effects Fisher had hoped for?" A close textual examination of the reviews of *The Genetical Theory of Natural Selection* suggests that, for the most part, Fisher's efforts to argue for the practical virtue of mathematical biology and establish goodwill for his enterprise generally went unnoticed by reviewers. Out of the eight reviews examined, only two made any comment on his ethical appeals, and of those, only one seriously addresses them.[12]

Although there was generally a lack of response to the ethical strategies in Fisher's work, the one reviewer who did comment on them, the geneticist R.C. Punnett, was one of the two reviewers who had dismissed Fisher's first paper, "The Correlation between Relatives." His review of the text offers an interesting perspective on the effect of Fisher's rhetorical efforts because it suggests that after twenty years of trying, Fisher had managed to make some small headway with an early critic of his work. A close reading of Punnett's review suggests that, (1) he was interested in the relationship between the mathematical, the experimental, and the observational components of genetics, (2) he had been generally predisposed to reject the importance of mathematics to biological investigations, and (3) Fisher had managed to make a sufficiently compelling case for the importance of mathematics to biological research and that Punnett was willing to accept his position on the issue.

The degree of importance Punnett ascribes to the ethical arguments in Fisher's work is evidenced by the location of the topic in his review and the amount of attention he devotes to it. Unlike other reviewers, who began their discussion of *The Genetical Theory of Natural Selection* by commenting on Fisher's argument for the importance of Darwin's theory of natural selection, Punnett opens his review with a reflection on Fisher's rhetorical argument.

From the very first line of his review, Punnett articulates the reluctance that he and other geneticists share about the benefit of collaborating with mathematicians: "Probably most geneticists today are somewhat skeptical as to the value of the mathematical treatment of their problems" (595). In the context of this general skepticism, Punnett shares with the readers of *Nature* his interpretation of Fisher's general arguments in the preface of the book. In his response to Fisher's claim that the difference between the mind of the biologist and the mathematician is only the result of a difference in training, Punnett voices his skepticism:

> In spite of Dr. Fisher's view, it is not unlikely that there may be real genetical difference in the types of mind respectively associated with biological and mathematical thought, so that the matter-of-fact intelligence of the former will seldom be in a position to make much of a response to the imaginative flights of the latter. (595)

The last line of this statement reveals Punnett's hostility to Fisher's general assertion that the difference between biological and mathematical thought was one of scope rather than kind. Punnett maintains the pragmatic critique by characterizing biology as being "matter-of-fact intelligence," or empirical knowledge with the virtue of practicality, while portraying mathematical analysis as typified by "imaginative flights" of reason.

Despite Punnett's general skepticism about the benefits of mathematical reasoning to biological research, he maintains that Fisher's specific efforts to show the rewards of symbiotic engagement between the fields are compelling. His conviction rests squarely on the statistical biologist's attempts to establish a positive ethos for his mathematical program of biology:

> Nevertheless, it is at times worth the biologist's while to make a special effort, and the present volume offers an occasion; for although Dr. Fisher's mind is essentially a mathematical one, he has marked biological sympathies, and has evidently striven hard to make himself comprehensible to those without mathematical *flair.* (595)

In these lines Punnett suggests that the book is relevant and can speak to the interests of genetic researchers on two grounds. First, he suggests that *The Genetical Theory of Natural Selection* is a legitimate, symbiotic endeavor between mathematics and science because Fisher establishes a credible ethos as a mathematical researcher with "biological sympathies." Punnett's comment here suggests that Fisher's concerted effort to forge an identification between himself and his biological audience by using Mendel's work as the basis for his mathematical argument, emphasizing the practical and empirical qualities of biomathematical research, had some positive effect on geneticists. Through his rhetorical overtures, Fisher established himself as an honest broker between the fields of mathematics and science in a way that Pearson, who regularly advocated the superiority of mathematics over science, was never able to accomplish.

In addition to winning the respect of biological researchers by identifying with their beliefs and practices, Punnett argues that Fisher invites cooperation by making his mathematical reasoning accessible to the biologists. This recognition suggests that Fisher's strategic efforts to accommodate his non-mathematical audience by limiting the extent and complexity of the mathematics in the book, and by offering natural language descriptions of his mathematical procedures, had made them more accessible. Though Punnett ends his review with a scathing diatribe on the lucidity of Fisher's prose style, and Fisher responds angrily and publically in *Nature* to some points of Punnett's critique, an understanding remains between the two that a small but significant détente has been reached between the conventional geneticists and the mathematical biologists.

Conclusion

R.A. Fisher's work has long been of interest to philosophers and historians of science because of its instrumental role in the development of the modern field of population genetics. Previous efforts have endeavored to explain its general success in terms of a conceptual capacity to overcome the divisive ideological issue of continuous and discontinuous variation, or to offer his audience the promise of future breakthroughs in the study of variation, evolution, and heredity (Provine; Plutynski). This chapter has endeavored to examine the unexplored possibility that Fisher's contribution to developing the field of population genetics was just as much rhetorical as it was philosophical or analytical.

An examination of the state of biometry after its unsuccessful challenge of Mendelism in the first decade of the twentieth century suggests that, to win supporters amongst conventional biologists for a mathematical program for the study of variation, evolution, and heredity, Fisher had to overcome substantial ethical obstacles. Close textual analysis of Fisher's earliest work, "The Correlation between Relatives," and the responses it drew from critiques reveal that, at least initially, Fisher was not fully conscious of the ethical challenges he had to overcome. Because of his initial failures and subsequent experience at Rothamsted, however, he began to understand the importance of making his work accessible and relevant to his audience. Evidence from later work, in "On the Dominance Ratio," reveals that Fisher had not only begun to recognize these persuasive obstacles, but had also developed strategies to overcome them by accommodating his work and actively arguing for the practical virtues of his methodology.

In *The Genetical Theory of Natural Selection* we witness the culmination of Fisher's efforts to establish a positive ethos for his program of mathematical investigation. In the book he displays his rhetorical savvy by choosing a title to attract geneticists and by making the mathematics in the text as accessible as possible. In addition, he attempts to change the conceptual frame that regularly undermines the ethos of mathematical biology by confronting critiques of its abstractness and lack of practicality. Although his arguments in the "Preface" of *The Genetical Theory of Natural Selection* were generally not recognized as contributing to a change in his readers' minds about the virtue of mathematics, his accommodation of mathematical arguments and efforts to identify his work with the interests of biologists succeed in persuading R.C. Punnett, an influential geneticist and early critic of his work, to admit that there might be something of value in this new path of exploration.

Afterword

In this book I have attempted to trace the emergence of mathematical argument in the study of variation, evolution, and heredity from Darwin and Mendel, through the biometricians and their debates with the Mendelians, to the efforts of Fisher. Whereas others have explored the shifting path of ideas that accompany the emergence of modern mathematical approaches to biology such as population genetics, I have focused my attention on argument, particularly (1) the conventions for making mathematical arguments in science, (2) the strategies by which novel applications of mathematical formulae and principles were argued for, and (3) whether those strategies were successful.

This overview of the conventions and strategies, and the successes and failures of those strategies suggests that the process by which modern mathematical investigations of variation, evolution, and heredity developed was neither simple nor straightforward. Instead, their development was slow and often rhetorical strategies to move forward. Close examinations of the texts and contexts of arguments made by Darwin, Mendel, Galton, Pearson, and Fisher have revealed the rhetorical dimensions of the growth of mathematical investigations of variation, evolution, and heredity. They have illustrated how mathematics can be a source of invention about nature and shown that mathematics can function as a value upon which the ethos of a scientific argument can be established, or a disputed value in a hierarchy of values. In addition, they have revealed how mathematical warrants themselves can rely on rhetorical appeals for their support, depending on arguments from ethos, narratives, figures of speech, visuals, and socio-cultural values to persuade audiences to accept them.

By illuminating these rhetorical dimensions of mathematics in science, a new picture of science and mathematics emerges, which suggests that empirical evidence, deductive reasoning, and other special lines of argument commonly co-exist with commonly used rhetorical ones. In this new vision, mathematics that participates in argument outside of its own domain can no longer be construed as off-limits or uninteresting for rhetoricians. The value of mathematics both as a rhetorical strategy and as a feature that can be argued for rhetorically suggests that the use of mathematics demands the attention of rhetoricians. Additionally, the fact that members of specific scientific discourse communities recruit general lines of reasoning to support mathematical arguments in their particular communities challenges the notion of a strict demarcation between the rhetorical and scientific argument. Instead of clear boundaries between specialist and generalist arguments, there seems to be a nested, overlapped, and reticulated system of argument practices. By exposing this interconnected and multifaceted system of argument, this book has endeavored to show that the intertwining of the rhetorical, analytical, and empirical is more substantial than some rhetorical analysts and theorists have allowed, and the methods of telling them apart are currently underdeveloped.

The present lack of study of the rhetorical dimensions of mathematics and the paucity of understanding of the nature and extent of the interconnectedness of discourse communities presents both problems and opportunities for the rhetorical analyst. I hope this book inspires others to contribute their expertise and insight to identifying and filling the many lacunae about the rhetorical dimensions of mathematics in discourse and argument.

Appendix A

List of Scholarly Books Analyzed for Citations of the Role of Mathematics in Darwin's Arguments

Aydon, Cyril. *Charles Darwin: The Naturalist Who Started a Scientific Revolution*. New York: Carroll and Graf, 2002. Print.

Aydon doesn't mention mathematics in the index of his text. His only reference is in his comment that Darwin was not particularly good in mathematics in school (30).

Bowlby, John. *Charles Darwin: A New Life*. New York: Norton, 1991. Print.

Bowlby doesn't mention mathematics in the index of his text. Chapter 22, "My Abominable Volume," discusses generally the contents of the argument in Darwin's text. It does not, however, go into specifics. Also, it does not include any mention of mathematics.

Bowler, Peter. *Charles Darwin: The Man and his Influence*. Cambridge: Cambridge UP, 1990. Print.

Bowler doesn't mention mathematics in the index of his text. He mentions that Darwin undertakes arithmetical studies of the number of species to prove that population pressure within a given area forces species to diversify in order to compete more successfully (104). He also discusses the importance of mathematical reasoning in Malthus's argument, which informs his concept of natural selection (117).

Brent, Peter. *Charles Darwin: A Man of Enlarged Curiosity.* New York: Harper and Row, 1981. Print.

Brent doesn't mention mathematics in the index of his text. In skimming chapters four and five, "Origin of all My Views" and "The Speculatist," I found no discussion of mathematics.

Browne, Janet. *Charles Darwin: The Power of Place.* Vol. 2. Princeton: Princeton UP, 1996. Print.

Browne doesn't mention mathematics in the index of her text. She talks at length in chapter two, "My Abominable Volume," about Darwin's use of evidence, but at no time discusses the role of mathematic notation in presenting that evidence.

Campbell, John Angus. "Charles Darwin: Rhetorician of Science." *The Rhetoric of the Human Sciences: Language and Argument in Scholarship and Public Affairs.* Ed. John S. Nelson. Madison: The U of Wisconsin P, 1987. 69–86. Print.

Campbell does not discuss the role of mathematics in his chapter of the text.

Gale, Berry G. *Evolution Without Evidence: Charles Darwin and the Origin of Species.* Albuquerque: U of New Mexico P, 1982. Print.

"Mathematics" is listed in the index of the text. Gale discusses Darwin's weakness in mathematics (13–14). He explains that Darwin read *Nature*, but that he did not understand most of the articles that used mathematics (79). Gale also discusses the contents of Darwin's arguments in depth. He looks at his arguments, their weaknesses, and the problems of evidence in chapter seven of the text. At no point does he discuss the role of mathematics in presenting Darwin's evidence.

Ghislen, Michael T. *The Triumph of the Darwinian Method.* Berkeley: U of California P, 1969. Print.

"Mathematics" is listed in the index of the text. Ghilsen defends Darwin's not using mathematics to make his geological arguments

(21). He argues that something can be logical whether it is expressed in a language of formal logic like mathematics or not (65). He attacks mathematical argument in the harder sciences (127). Ghilsen explains that Darwin uses geometry to describe changes in the physical shape of organisms. He also explains that Darwin quantifies different physical features in organisms to present precise, morphological comparisons (170). Finally, he discusses Darwin's hybridization experiments, claiming that his results were subjected to statistical analysis by Galton (175).

Gillespie, Neal C. *Charles Darwin and the Problem of Creation.* Chicago: U of Chicago P, 1979. Print.

Gillespie doesn't mention mathematics in the index. He doesn't discuss it with regard to Darwin's approach to his subject.

Hull, David. *Darwin and His Critics: The Reception of Darwin's Theory of Evolution by the Scientific Community.* Cambridge: Harvard UP, 1973. Print.

Hull has an entry for mathematics in the index with six page references, five of which are relevant to the relationship between Darwin, his work on evolution, and mathematics. In the first reference, Hull comments on how Darwin was confused about the differences between scientific laws, mathematical axioms, and metaphysical principles; however, he maintains that he was no more confused than most researchers in the period (11). In the second reference, he claims that, despite his misunderstanding of the differences between these categories of reasoning, Darwin understood the general distinction between deductive and inductive reasoning (14). The third reference to mathematics and Darwin is a discussion in which Hull explains that Darwin was caught in the middle of a debate about the difference between the nature of mathematical axioms and their relation to their experience. He argues that Darwin's theory was not mathematical in form or in the type of reasoning (32). In the next reference, Hull explains that, in the 1930s, different scientists tried, through genetics, to create a mathematical science of evolutionary biology. However, this was not effective in creating verifiable outcomes (34). In the final reference,

Hull concludes that evolution did not exhibit mathematical regularity as did the laws in hard science (61).

Mayr, Ernst. *Charles Darwin and the Genesis of Modern Evolutionary Thought*. Cambridge: Harvard UP, 1991. Print.

Mayr doesn't mention mathematics in the index of his text. Mayr notes that the concept of science in Darwin's time was completely dominated by mathematics and the physical sciences (48).

Appendix B

Reviews of William Bateson's *Mendel's Principles of Heredity* (1909)

Positive reviews have been have been marked with a (+)

(+) "Future Generations." Rev. of *Mendel's Principles of Heredity. Observer* 13 June 1909. Print.

(+) "Genetics Made Easy." Rev. of *Mendel's Principles of Heredity. Country Gentleman* 8 May 1909. Print.

(+) "Mendel and Heredity." Rev. of *Mendel's Principles of Heredity. Tablet* 19 June 1909. Print.

Rev. of *Mendel's Principles of Heredity. Athenaeum* 4 Sep. 1909. Print.

(+) —. *Contemporary Review* June 1909. Print.

—. *Daily Telegraph* 3 June 1909. Print.

—. *Glasgow Herald* 21 May 1909. Print.

(+) —. *Knowledge* Jan. 1909. Print.

—. *Lancet* 173 4473 1909 pgs. 1461–62. Print.

—. *Literary World* June 1909. Print.

—. *Liverpool Daily Post* 7 May 1909. Print.

(+) —. *Livestock Journal* 7 May 1909. Print.

—. *Manchester Courier* 28 May 1909. Print.

(+) —.*Manchester Guardian* 29 Apr. 1909. Print.

(+) —. *Nation* 16 Sep. 1909. Print.

—. *Saturday Review* 22 Jan. 1910. Print.

(+) —. *Scotsman* 29 Apr. 1909. Print.

(+) —. *Sheffield Telegraph* 29 Apr. 1909. Print.

(+) —. *Standard* 27 Apr. 1909. Print.

(+) "The New Science." Rev. of *Mendel's Principles of Heredity. Birmingham Post* 7 May 1909. Print.

(+) Wood, Eugene. "A Mendelian Review." Rev. of *Mendel's Principles of Heredity. Wilshire's* Feb. 1910. Print.

Notes

Introduction

1. For Kant, however, the term "science" had a different meaning that we ascribe to it today. For him, "science" was a general label for rigorous knowledge/argument claims, whereas today we use it most commonly to describe a particular field of knowledge production.

2. Investigations of the mathematical dimensions of scientific argument have been undertaken by Ashmore ; Cifoletti; Dear; Gross, Harmon, and Reidy; Hunt; Miller; and Reyes.

3. Prelli discusses the scientific *topoi* in Chapter 9 of the text. The *problem-solution topoi* seem to have their source in Kuhn's discussion of scientific paradigms in Chapter 4 of *The Structure*, "Normal Science as Puzzle Solving." Prelli's *evaluative topoi* coincide with the reasons that can be used to argue in favor of a paradigm during a revolutionary period. These appear in Chapter 12 of *The Structure*, "The Resolution of Revolutions" (153–58).

4. This methodological problem is discussed at length in Paul, Charney, and Kendall.

5. Fisher's *The Genetical Theory of Natural Selection* was published before the advent of the field population genetics, which it helped to found. The text was published in 1929, and the first recorded reference to a field by the name of "population genetics" was not made until 1938, according to the OED.

Chapter 2

1. See *The Controversy on the Comets of 1618: Galileo Galilei, Mario Guiducci, and Johann Kepler*, pages 183–84.

2. Both Whewell and Herschel were, for example, founding members of the British Association for the Advancement of Science (BAAS).

3. The Tripos was an important test that was the basis for honors at Cambridge. Before 1848, the only way to obtain honors at Cambridge was through the Mathematical Tripos. See Fisch (24).

4. For a discussion of these debates, see Snyder and Fisch.

5. See Whewell, *Philosophy* 1: 84.

6. Neither Herschel nor Whewell use the phrase, "quantitative inductive process." I am employing it as a way to differentiate between induction that is dedicated to determining causes, which Whewell calls "the induction of causes" (*Philosophy* 2: 431), and induction devoted to the development of empirical laws from quantified data.

7. Whewell capitalizes the term "Idea" throughout the book to separate the fundamental cognitive principles to which he is referring from "ideas," which are the stuff of everyday, mental activity. In order to maintain this distinction, I will also capitalize the term.

8. Later, in Volume II, he also introduces "resemblance" as a primary quantitative Idea based on the axioms of arithmetic (*Philosophy* 2: 412–25). This stipulation is important because there were other types of Ideas, such as substance, force, and polarity, that were qualities thought not to be amenable to quantification because no method or scale for their measurement had been conceived.

9. $2 * 1 + 1(1 * 1) = 3$; $2 * 2 + 1(2 * 2) = 8$; $2 * 3 + 1(3 * 3) = 15$

10. In its very narrowest technical sense, analogy is a comparison of two relationships: *A* is to *B* as *C* is to *D*. In mathematical scientific argument, it is not always the case that there are two features whose relationships are being compared. As a result, I am defining analogy in more broadly.

11. The cited material is taken from *The Complete Works of Aristotle.*

12. By "vaguely understood," I mean here that we have traces of the underlying law dictating the change provided by the data, but no firm understanding of what that law may be.

Chapter 3

1. The material in the *Variation of Plants and Animals under Domestication* comes from the first two chapters of the "big species book" manuscript.

2. Life insurance companies, which emerged in the middle of the eighteenth century, were particularly interested in vital statistics. They used them to produce life tables for calculating annuities. See Cullen 7–8.

3. This final question is a direct quote from Whewell's *History of the Inductive Sciences*, 475.

4. This source and the lines excerpted from it can be found in Robert Brown, *The Miscellaneous Works of Robert Brown.*

5. Monocotyledons and dicotyledons are the two major classes of flowering plants. Monocotyledons are so named because they have a single cotyledon (or embryonic leaf) in their seeds, whereas dicotyledons have two. ("Monocotyledon")

6. All citations from Darwin's notebooks come from *Charles Darwin's Notebooks* but are labeled here according to the notebook in which the cited text appears.

7. This source and the lines excerpted from it come from *The Complete Works of Aristotle.*

8. In a memorandum in the *Natural Selection* manuscript, there is a note in which Darwin identifies his primary audience as "geologists and zoologists" (94). Given that many of the examples and arguments in the text are drawn from botany, I am also assuming that botanists were also considered by Darwin as a primary audience.

9. Arguments for natural selection and the struggle for existence are made in chapter three of *The Origin of the Species*, "Struggle for Existence."

10. In this letter, Darwin explains to Lyell that this is how Herschel's reaction to the law had been described.

11. In a reply to the philosopher John Stuart Mill, Sedgwick wrote that Darwin's theory in The *Origin of the Species* "is like a pyramid poised on its apex. It is a system embracing all living nature . . . yet contradicting point blank the vast treasure of facts that the Author of Nature has, during the past two or three thousand years, revealed to our senses" (qtd. in Hull 169).

Chapter 4

1. The timeframe of Mendel's experiments vary in different historical accounts. They can start as early as 1854 and end as late as 1865. I have chosen these dates based on majority opinion (Henig, Iltis, Olby, and Orel).

2. Rudolph Jacob Camerer is better known under his Latinized name, Camerarius.

3. Neither Kölreuter nor Gärtner had a theory that included the genotype, the collection of both expressed and hidden characters.

4. Ich erzog hievon drey Pflanzen. Eine derselben war ihrer ganzen äusserlichen Anlage nach dem in der zweyt. Forts. § 16. S. 73 etc. besschriebenen Bastart im ersten aufsteigenden Grade sehr ähnlich, und hinterliess viele, aber ganz leere Kapseln. (From this experiment I grew three plants. One of those was in its whole external structure very similar to the previously described (second continuation, paragraph 16, p. 73 etc) hybrid of the first increasing degree, and left many, but quite empty, seed pods.) Kölreuter defines "first increasing degree" in the *Second Continuation* as when a hybrid produces offspring after breeding with plants of one of the original parent species, which maintains the hybrid characteristics that are foreign to either parent. He writes, "Jene nenne isch Bastarte im absteigenden Grade, weil sie einen Theil ihrer fremden Gestalt abgelegt, statt dessen aber von ihrer eigenen wieder etwas angenommen haben, so, dass nun ihre eigenthümilche Natur die Oberherrschaft über die fremde bekommen hat: diese hingegen

nenne ich Bastarte im aufsteigenden Grade, weil bey ihnen gerade das Gegentheil von dem, was bey jenen vorgegangen, geschehen ist." (Those I call hybrids in decreasing degree, because they discarded a part of their foreign [hybrid] form, but instead re-adopted some of their own [previous character from the parents], so that now their original nature became dominant over the foreign one: on the other hand, these I call hybrids in increasing degree, because exactly the opposite of that, which happened in those [hybrids of decreasing degree], happened in them.) (*Vorläufige Nachricht* 196, 72).

5. Die zwo übrigen hatten etwas weniger Aehnlichkeit mit der panic. Als die erstern, und setzennur sher wenige, ziemlich spitzige und ebenfalls ganz leere Kapseln an. (The other two had somewhat less similarity with the *panic.* than the first, and produced only very few, rather pointed and also quite empty seed pods.) (*Vorläufige Nachricht* 196).

6. See also Iltis, 106–107.

7. Es gehört allerdings einiger Muth dazu, sich einer so weit reichenden Arbeit zu unterziehen; indessen scheint es der einzig richtige Weg zu sein, auf dem endlich die Lösung einer Frage erricht warden kann, welche fur die Entwicklungs-Geschichte der organischen Formen von nicht zu unterschätzender Bedeutung ist. (*Versuche* 4). The original German is from a reprint of Mendel's original text: Gregor Mendel, *Versuche über Plfanzen-Hybriden* edited by J. Cramer and H.K. Swann. All subsequent footnotes containing the original German will be from this text.

8. Allein nichts berechtigt uns zu der Annahme, das die Neigung zur Varietätenbildung so ausserordentlich gesteigert werde, dass die Arten bald alle Selbstständigkeit verlieren un ihre Nachkommen in einer endlosen Reihe höchst veranderlicher Formen auseinander gehen. (*Versuche* 36).

9. Die aufallende Regelmässigkeit, mit welcher dieselben Hybridformen immer wiederkehrten, so oft die Befruchtung zwischen gleichen Arten geschah, gab die Anregung zu weiteren Experimenten, deren Aufgabe es war, die Entwicklung der Hybriden in ihren Nachkommen zu verfolgen. / Dieser Aufgabe haben sorgfältige Beobachter, wie Kölreuter, Gärtner, Herbert, Lecocq, Wichura u.a. einen Theil ihres Lebens mit unermudlicher Ausdauer geopfert. (*Versuche* 3).

10. A German translation of *The Origin of Species* was prepared by Heinrich Bronn, a German paleontologist, and released in 1860 only a few months after the release of the English version (Gliboff 4). Of Darwin, Mendel has been reported as saying that he was, "greatly interested in the ideas of evolution, and was far from being an adversary of the Darwinian theory." However, he is also reported to have stated that, "there was still something lacking" (Orel 71).

11. Orel does not provide the name of the textbook. I assume, however, that he is referring to *Arithmetik und Algebra: mit besonderer Rücksicht auf*

die Bedürfnisse des practischen Lebens und der technischen Wissenschaften: nebst einem Anhange von 450 Aufgaben, published in 1844.

12. Die Verhältnisse, nach welchen sich die Abkömmlinge der Hybriden in der ersten und zweiten Generation entwickeln und theilen, gelten wahrscheinlich für alle weiteren Geschlechter. (*Versuche* 17).

13. Bezeichnet *A* das eine der beiden constanten Merkmale, z. B. das dominirende, *a* das recessive, und *Aa* die Hybridform, in welcher beide vereingt sind, so ergibt der Ausdruk:

$$A + 2Aa + a$$

die Entwicklungsreihe für die Nachkommen der Hybriden je zweier differirender Merkmale. (*Versuche* 17).

14. Die von Gärtner, Kölreuter und Anderen gemachte Wahrnehmung, dass Hybriden die Neigung besitzen zu den Stammarten zurückzukehren, ist auch durch die besprochenen Versuche bestätigt. . . . Nimmt man durchschnittlich für alle Pflanzen in allen Generationen eine gleich grosse Fruchtbarkeit an, erwägt man ferner, dass jede Hyride Samen bildet, aus denen zur Hälfte wieder Hybriden hervorgehen, während die andere Hälfte mit beiden Merkmalen zu gleichen Theilen constant wird, so ergeben sich die Zahlenverhältnisse für die Nachkommen in jeder Generation aus folgender Zusammenstellung. (*Versuche* 17–18).

15. Erster Versuch:

AB samenpflanze, *ab* Pollenpflanze,
A Gestalt rund, *a* Gestalt kantig,
B Albumen gelb, *b* Albumen grün. (*Versuche* 19)

16. 315 rund und gelb.
101 kantig und gelb,
108 rund und grün,
32 kantig und grün. (*Versuche* 19).

17. 38 runde gelbe Samen *AB*
65 runde gelbe und grün Samen *ABb*
60 runde gelbe und kantig gelbe Samen *AaB*
138 runde gelbe und grüne, kantige gelbe und grüne Samen *AaBb* (*Versuche* 19).

18. 38 Pflanzen mit der Bezeichnung (*Versuche* 20).

19. Daher entwickeln sich die Nachkommen der Hybriden, wenn in denselben zweierlei differirende Merkmale verbunden sind, nach dem Ausdrucke. (*Versuche* 20).

20. Diese Entwicklungsreihe ist unbestritten eine Combinationsreihe, in welcher die beiden Entwicklungsreihen für die Merkmale *A* und *a, B* und

b gliedweise verbunden sind. Man erhält die Glieder der Reihe vollzählig durch die Combinirung der Ausdrücke. (*Versuche* 21).

21. Es unterliegt daher keinem Zweifel, dass für sämmtliche in die Versuche aufgenommenen Merkmale der Satz Giltigkeit habe: die Machkomme der Hybriden, in welchen mehrere wesentlich verschienden Merkmale vereinigt sind, stellen die Glieder einer Combinationsreihe vor, in welchen die Entwicklungsreihen für je zwei differende Merkmale verbunden sind. (*Versuche* 22).

22. Verhalten je zweier differirender Merkmale in hybrider Verbindung unadhängig ist von den anderweitigen unterschieden an den beiden Stammpflanzen. (*Versuche* 22).

23. If, for example, there are two different characters pairs, *Aa* and *Bb* (n=2), then there should be nine (3^2) possible combination possibilities, *AB*+*ab*+*Ab*+*aB*+*2AaB*+*2Aab*+*2ABb*+*2aBb*+*4AaBb*, which describe sixteen different individuals (4^2) and have four (2^2) constant unions, *AB, ab, Ab, and aB.*

24. Das Gesetz der Combinirung der differirenden Merkmale. (*Versuche* 32).

25. Bezeichnet *n* die Anzahl der characteristischen Unter schieden an den beiden Stammpflanzen, so gibt 3^n die Gliederzahl der Combinationsreihe, 4^n die Anzahl der Individuen, welchein die Reihe gehören, und 2^n die Zahl der Verbindungen, welche constant bleiben. (*Versuche* 22–23).

26. Dass constante Merkmale, welche an verschiedenen Formen einer Pflanzsippe vorkommen, auf dem Wege der weiderholten künstlichen Befruchtung in alle Verbindungen treten können, welche nach den Regeln der Combination möglich sind. (*Versuche* 23).

27. Es bleibt ganz dem Zuffalle überlassen, welche von den beiden Pollenarten sich mit jeder einzelen Keimzelle verbindet. Indessen wird es nach den Regeln der Wahrscheinlichkeit im Durschnitte vieler Fälle immer geschehen, dass sich jede Pollenform *A* und *a* gleich oft mit jeder Keimzellform *A* und *a*. (*Versuche* 29).

28. In the front matter of the *Proceedings*, there are 115 organizations listed with whom the Brünn Natural Society had communications with in the year 1865. This list may offer a clue to the scope of the circulation of the journal, and, therefore, Mendel's ideas in broader scientific circles. See *Verhandlungen des naturforschenden Vereines in Brunn*.

Chapter 5

1. See Campanario; Paul; Gieryn; Prelli, "The Rhetorical Construction"; and Myers.

2. *Traditional probability* relies on past outcomes to calculate the probability of future outcomes, while *subjective probability* makes guesses about

the probability of future outcome for events that have never happened in the past.

3. Donald Mackenzie offers a compelling explanation of why the applications for comparison, though mathematically available, did not exist in a conceptual framework that suggested their application in the social and biological sciences (68–72).

4. Galton describes the Ancestral Law of Heredity in the article, "Family Likeness in Eye Color" (*Nature* 137), and gives a more detailed, mathematical treatment in an article of the same title in the *Proceedings of the Royal Society* (402–416).

5. Regression is explored in Galton's article "Hereditary Stature" in *Nature* and in "Regression towards Mediocrity in Hereditary Stature" in the *Journal of the Anthropological Institute.*

6. In *Hereditary Genius,* for example, he writes "I acknowledge freely the great power of education and social influences in developing the active powers of the mind, just as I acknowledge the effect of use in developing the muscles of a blacksmith's arm, and no further. . . . There is definitely a limit to the muscular powers of every man, which he cannot by any education or exertion overpass" (14–15).

7. For Galton, the term "scheme" refers to, "a compendium of a mass of observations which, on being marshaled in an orderly manner, fall into a diagram whose contour is so regular, simple, and bold, as to admit of being described by a few numerals . . . from which it can at any time be drawn afresh" (49).

8. See Dewey and Venn and reviews of *Natural Inheritance* in *Science, Times*, and *Nation.*

Chapter 6

1. The other measurements include "the distance from the posterior margin of the carapace to the front of the median spine; the length of the sixth abdominal tergum [the small plates on the back of the shrimp]; and the length of the telson [the end spike which protrudes over the tail fan of decapods]" (Weldon, *Cragon Vulgaris* 448).

2. These measurements were gathered at Plymouth on the southern coast, Southport on the western coast, and Sheerness on the eastern coast.

3. Herbert Thompson worked at the zoological laboratory at University of College London. On suggestion from Weldon, he gathered measurements of 1,000 prawns. Pearson uses these measurements in his papers on frequency curves from variation data. He published the results of his research in the *Proceedings of the Royal Society London* under the title, "On the Correlation of Certain External Parts of Paloemon Serratus" (1894).

4. Though Bateson makes the case that discontinuous variation exists and that it plays the seminal role in speciation, he does not dismiss the existence of continuous variation.

5. Galton's placation strategy is discussed in Rosemary Harvey's *William Bateson and the Emergence of Genetics* (35), and verified by the letters from Weldon to Galton of November 17th and 26th, 1896 (Galton Collection University College London). Hybridization experiments are discussed in letters from Bateson to Galton exchanged on December 3rd and 24th, 1896.

6. The letters between Weldon and Galton and Weldon and Pearson make no mention of personal conflict with Bateson or the work of the Evolutionary Committee during the period between 1897and 1899.

7. Pearson uses the basic equation for arriving at the number of possible pairs, if order of the pairs doesn't matter: n (n-1)/2. Because there are twenty-six leaves from each tree (n=26), such that we arrive at the equation (26 x 25) ½, which is equal to 325.

8. Pearson uses the term "bathmic" here to refer to an "inherent growth force" which is responsible for the variation in individual cells. He offers an extended discussion of this concept in *The Grammar of Science* (1900) (375–77).

Chapter 7

1. See the reviews of *Mendel's Principles of Inheritance* in *Athenaeum*, *Daily Telegraph*, *Lancet*, and *Saturday Review*.

2. According to Galton's Law of Ancestral Hereditary, inheritance was the sum of the variability of a particular trait that an individual received from their immediate parents and their more remote ancestors. Based on his calculations of regression using data on stature, Galton concluded that offspring deviate from the mean of the population for a particular character two thirds as much as their parents He also argued that parents tended to deviate from the population mean for a particular character by only one third as much as their offspring, meaning that it was more likely that an exceptional offspring would arise from normal parents than be the consequence of a slight diminution of variability in exceptional parents (*Natural Inheritance* 98–100).

Using the law of regression as a baseline, Galton calculated the amount of variation that an individual could expect to inherit for a particular trait from their total ancestry. Assuming that parents always deviated from the mean by only one third as much for a particular character as their offspring, and assuming that the contribution of variation from ones ancestry diminished geometrically with each preceding generation, he concluded that the total bequeathal of variation in any particular trait was 3/2:

> Call the peculiarity [value of variation from the population mean] of the Mid-Parent D, then the implied peculiarity of the Mid-parent of the mid-parent, that is the mid-grand-parent of the man, would on the above supposition be 1/3D, that of the mid-great-grand-parent would be 1/9D, and so on. Hence the total bequeathable property would amount to D (1+1/3+1/9+&c.) = D3/2. (Galton, Natural Inheritance, 134)

The value of 3/2 for heritable variation, however, was much higher than the regression value of two thirds that Galton had calculated when comparing the mean of the stature of parents and offspring. As a consequence, Galton concluded that there must be forces at work which diminish variation. He does not speculate on what these forces are; however, he calculates their value. Assuming that these forces are at work to an equal extent in every generation, he calculates that for the deduced value for heritable variation 3/2 to be diminished to the measured value of two thirds, there would have to be reduced by 4/9, or roughly one half. As a consequence, inheritance would diminish geometrically by one half for every generation we follow a character back: "Hence the influence . . . of the mid-parent may be taken as 1/2 and that of the mid-grandparent as 1/4 and so on" (135–36).

3. In Yule's case, this compatibility was limited to crosses amongst dominant characters. In Pearson's it was only when inheritance was considered under Mendel's original law (i.e., barring all of the new biological phenomena such as incomplete dominance, polymorphism, epistacy, etc.).

4. Later in the paper Fisher cites both the papers by Pearson and Yule discussed earlier in this section.

5. Epistacy is the interaction of two traits to create joint effects. Polymorphism is when more than one allele or genetic pair of traits exists, usually for a particular physically expressed characteristic in a population.

6. Fisher calculated the fraternal correlation to be .5 and the fraternal parent correlation to be .25 ("Correlation" 405).

7. Fahnestock opens the article by writing, "Two thousand years ago, Crassius speaking for Cicero in the dialogue *De Oratore*, could have been describing those public intermediates, the orators of magazines and newspaper columns who interpret the wonders of twentieth century science for lay readers" (330–31).

8. In the citations of the three texts independent of the author or his sponsor, "The Correlations between Relatives" is referenced once in each text. The references are to specific ideas defined in Fisher's work rather than to the larger implications of the work either as a bridge between biometry and genetics or as a proof of the limited effect of the environment on hereditary outcomes.

9. See Leonard Darwin, "Eugenics as Related to Economics and Statistics," as well as Fisher, "Triplet Children in Great Britain and Ireland."

10. These conflicts are said to have developed over a critique of Fisher's ideas in "Frequency Distribution of the values of the Correlation Coefficient in Samples of an Indefinitely Large Population," by Pearson and other members of the Galton Laboratory in their paper, "On the Distribution of the Correlation Coefficient in Small Samples." Fisher complained and was later vindicated in his complaint that the critique was unjustified because it resulted from a fundamental misunderstanding of his statistical model. A discussion of the conflict is presented by Joan Fisher Box in her biography of her father (70–88), and responded to by Karl Pearson's son Egon Sharpe (E.S.) Pearson in *Biometrika* (55.3, 1968).

11. In a letter dated May 28, 1929, Sisam writes to Fisher: "In its present form the book would be very useful to specialists; but—though it could never be 'popular'—the circle of readers could be increased considerably (especially in America) if the treatment were simplified in hard places, so as to bring more of it in reach of those not highly equipped already" (qtd. in Bennett p. 20).

12. The reviews of *The Genetical Theory Examined* include: "Biological Theory;" "Problems of Life;" Review of *The Genetical Theory of Natural Selection*, *Quarterly Review of Biology*; Charles Galton Darwin; Haldane; Hill; Punnett; and Wright. Only "Biological Theory" and Punnett comment on the ethical appeals.

Works Cited

Abrams, Philip. *The Origins of British Sociology: 1834–1914.* Chicago: U of Chicago P, 1968. Print.

Aristotle. *The Basic Works of Aristotle.* Ed. Richard McKeon. New York: Random House, 1941. Print.

—.*The Complete Works of Aristotle.* Ed. Jonathan Barnes. Vol. 2. Princeton: Princeton UP, 1984. Print.

—. *Rhetoric.* Trans. W. Rhys Roberts. New York: Random House, 1984. Print.

Ashmore, Malcolm. "Fraud by Numbers: Quantitative Rhetoric in the Piltdown ForgeryDiscovery." *Mathematics, Science, and Postclassical Theory.* Eds. Barbara Hernstein Smith and Arkady Plotnitsky. Durham: Duke UP, 1997. Print.

Bacon, Francis. *Novum Organum.* Trans. and Eds. Peter Urbach and John Gibson. Chicago: Open Court, 1994. Print.

Bateson, Anna, and William Bateson. "On the Variations in the Floral Symmetry of Certain Plants Having Irregular Corollas." *Journal of the Linnaean Society* 28.196 (1891): 386–424. Print.

Bateson, William. "To Francis Galton." 3 Dec. 1896. TS. The Bateson Letters Collection. John Innes Horticultural Institute, Norwich.

—. "To Francis Galton." 24 Dec. 1896. TS. The Bateson Letters Collection. John Innes Horticultural Institute, Norwich.

—. "Heredity, Differentiation, and other Conceptions of Biology: A Consideration of Professor Karl Pearson's Paper 'On the Principle of Homotyposis.'" *Proceedings of the Royal Society of London* 69 (1901–1902): 193–205. Print.

—. *Mendel's Principles of Heredity: A Defense.* Cambridge: Cambridge UP, 1902. Print.

—. *Materials for the Study of Variation.* London:Macmillan and Co., 1894. Print.

Beer, Gillian. Introduction. *The Origin of Species.* By Charles Darwin. Oxford: Oxford UP, 1996. vii-xxviii. Print.

Bennett, J.H. ed. *Natural Selection, Heredity, and Eugenics: Including Selected Correspondences of R.A. Fisher with Leonard Darwin and Others.* Oxford: Clarendon, 1983. Print.

Berkenkotter, Carol, and Thomas Huckin. *Genre Knowledge in Disciplinary Communication Cognition / Culture / Power.* Hillsdale: Erlbaum, 1995. Print.

"Biological Theory." Rev. of *The Genetical Theory of Natural Selection. Times Literary Supplement* 28 Aug. 1930. 677. Print.

Bowler, Peter. *Charles Darwin: The Man and His Influence.* Cambridge: Cambridge UP, 1990. Print.

Box, Joan Fisher. *R.A. Fisher: The Life of a Scientist.* New York: Wiley, 1978. Print.

Brookes, Martin. *Extreme Measures.* New York: Bloomsbury, 2004. Print.

Brown, Robert. *The Miscellaneous Botanical Works of Robert Brown.* Ed. John J.Bennett. Vol. 1. London, 1866. *Google Book Search.* Web. 12 June 2008.

Browne, Janet. "Darwin's Botanical Arithmetic and the 'Principle of Divergence,' 1854–1858." *Journal of the History of Biology* 13 (1980): 53–89. Print.

Bulmer, Michael. *Francis Galton: Pioneer of Heredity and Biometry.* Baltimore: Johns Hopkins UP, 2003. Print.

Burke, Kenneth. *A Rhetoric of Motives.*1950. Berkeley: U of California P, 1969. Print.

Camarer, Rudolph Jacob. Über das Geschlecht der Pflanzen. Leipzig, 1694. *Google Book Search.* Web. 22 Dec. 2009.

Campanario, Juan Miguel. "Peer Review for Journals as it Stands Today—Part 2." *Science Communication* 19.3 (1998): 277–306. Print.

Campbell, J. A. "Charles Darwin and the Crisis of Ecology: A Rhetorical Perspective." *Quarterly Journal of Speech* 60 (1974): 442–449. Print.

—. "The Polemical Mr. Darwin." *Quarterly Journal of Speech* 61 (1975): 375–390. Print.

—. "Charles Darwin: Rhetorician of Science." *The Rhetoric of the Human Sciences: Language and Argument in Scholarship and Public Affairs.* Ed. J. Nelson. Madison: U of Wisconsin P, 1987. Print.

—. "The Invisible Rhetorician: Charles Darwin's 'Third Party' Strategy." *Rhetorica* 7 (1989): 55–85. Print.

—. "Why was Darwin Believed?: Darwin's *Origin* and the Problem of Intellectual Revolution." *Configurations* 11.2 (2003): 203–37. Print.

de Candolle, Alphonse. "Essai Elementaire de Geographie Botanique." *Dictionnaire Classique D'Histoire Naturelle.* Paris, 1825. Print.

Cannon, Susan. *Science in Culture: The Early Victorian Period.* New York: Dawson, 1978. Print.

Canon, Walter. "John Herschel and the Idea of Science." *Journal of the History of Ideas* 22.2 (1961): 215–239. Print.

Chatterjee, Shoutir Kishore. *Statistical Thought: A Perspective and History.* Oxford: Oxford UP, 2003. Print.

Cicero. *De Inventione*. Trans. H.M. Hubbell. Cambridge: Harvard UP, 2000. Print.

Cifoletti, Giovanna. Introduction. "Mathematics and Rhetoric." *Early Science and Medicine* 11.4 (2006): 369–389. Print.

"Combinatorial." Def. 2. *Merriam-Webster's Collegiate Dictionary*. 10th ed. 1998. Print.

The Controversy on the Comets of 1618: Galileo Galilei, Mario Guiducci, Johann Kepler. Trans. Stillman Drake and C.D. O'Malley. Philadelphia: U of Pennsylvania P, 1960. Print.

Corcos, Alain, and Floyd Monaghan. *Gregor Mendel's Experiments on Plant Hybrids*. New Brunswick: Rutgers UP, 1966. Print.

Crowley, Sharon, and Deborah Hawhee. *Ancient Rhetorics for Contemporary Students*. 3rd ed. New York: Pearson Longman, 2004. Print.

Cullen, M. J. *The Statistical Movement in Early Victorian Britain*. Hassocks: Harvester, 1975. Print.

Darwin, Charles. *The Autobiography of Charles Darwin 1809–1882 with the Original Omissions Restored Edited and with Appendix and Notes by his Granddaughter Nora Barlow*. Nora Barlow Ed. London: Collins, 1958. Print.

—. *Charles Darwin's Natural Selection*. Ed. R.C. Stauffer. Cambridge: Cambridge UP, 1975. Print.

—. *Charles Darwin's Notebooks*. Ed. and Trans. Paul Barrett. Ithica: Cornell UP, 1987. Print.

—. "To Asa Gray." 24 Aug. 1855. *Darwin Correspondence Project*. American Council of Learned Societies and Cambridge University. Web. 17 July 2008.

—. "To Charles Lyell." 10 Dec. 1859. *Darwin Correspondence Project*. American Council of Learned Societies and Cambridge University. Web. 25 July 2008.

—. "To John Hooker." 22 Aug. 1857. *Darwin Correspondence Project*. American Council of Learned Societies and Cambridge University. Web. 1 July 2008.

—. "To John Lubbock." 14 July 1857. *Darwin Correspondence Project*. American Council of Learned Societies and Cambridge University. Web. 1 June 2010.

—. *The Origin of Species*.1859. Rpt. Oxford: Oxford UP, 1996. Print.

—. *The Variation of Animals and Plants under Domestication*. 2nd ed. 2 vols. New York: Appleton, 1883. Print.

Darwin, Charles Galton. "Natural Selection." Rev. of *The Genetical Theory of Natural Selection*. *The Eugenics Review* 22.2 (1930): 127–30. Print.

Darwin, Leonard. "Eugenics as Related to Economics and Statistics." *Journal of the Royal Statistical Society* 82.1 (1919): 1–33. Print.

Daston, Lorraine. *Classical Probability in the Enlightenment.* Princeton: Princeton UP, 1988. Print.

Davenport, Charles. "A History of the Development of the Quantitative Study of Heredity." *Science* 12.310 (1900): 864–870. Print.

Davis, Philip and Reuben Hersh. "Rhetoric and Mathematics." *The Rhetoric of the Human Sciences: Language and Argument in Scholarship and Public Affairs*. Ed. John S. Nelson. Madison: U of Wisconsin P, 1987. Print.

Dear, Peter. *Discipline and Experience: The Mathematical Way in the Scientific Revolution*. Chicago: U of Chicago P, 1995. Print.

Depew, David and Bruce Weber. *Darwinism Evolving: Systems Dynamics and the Genealogy of Natural Selection*. Cambridge: MIT P, 1995. Print.

Dewey, John. "Galton's Statistical Methods." Rev. of *Natural Inheritance. Publications of the American Statistical Association* 1 (1889): 331–34. Print.

Di Gregorio, Mario A. *Charles Darwin's Marginalia*. Vol. 1. New York: Garland, 1990. Print.

Doppler, Christian. *Arithmetik und Algebra: mit besonderer Rücksicht auf die Bedürfnisse des practischen Lebens und der technischen Wissenschaften: nebst einem Anhange von 450 Aufgaben*. Prague, 1844. Print.

Fahnestock, Jeanne. "Accommodating Science: The Rhetorical Life of Scientific Facts." *Written Communication* 15.3 (1998): 330–350. Print.

—. *Rhetorical Figures in Science*. New York: Oxford UP, 1999. Print.

—. "Series Reasoning in Scientific Argument: Incrementum and Gradatio and the Case of Darwin." *Rhetoric Society Quarterly* 26.4 (1996): 13–40. Print.

Feingold, Eleanor. Personal interview. 16 Nov. 2009.

Fisch, Menachem. *William Whewell: Philosopher of Science*. Oxford: Clarendon, 1991. Print.

Fisher, R.A. "The Correlation between Relatives on the Supposition of Mendelian Inheritance." *Transactions of the Royal Society of Edinburgh* 52 (1918): 399–433. Print.

—. "On the Dominance Ratio." *Proceedings of the Royal Society of Edinburgh* 42 (1922): 321–341. Print.

—. *The Genetical Theory of Natural Selection*. Oxford: Oxford UP, 1930. Print.

—. "Triplet Children in Great Britain and Ireland." *Proceedings of the Royal Society of London. Series B, Containing Papers of a Biological Character* 102 (1928): 286- 311. Print.

"From Father to Son." Rev. of *Natural Inheritance. New York Times* 14 Apr. 1889: 19. Print.

Gale, Berry G. *Evolution without Evidence: Charles Darwin and the Origin of Species*. Albuquerque: U of New Mexico P, 1982. Print.

Galison, Peter. "Ten Problems in History and Philosophy of Science." *ISIS* 99 (2008): 111–24. Print.

Galton, Francis. "Family Likeness in Eye Color." *Nature* 34 (1886): 137. Print.

—. "Family Likeness in Eye Color." *Proceedings of the Royal Society* 40 (1886): 402–16. Print.

—. *Hereditary Genius.* New York: St. Martin's, 1978. Print.

—. "Hereditary Stature." *Nature* 33.848 (1886): 295–298. Print.

—. "Letter to Karl Pearson." 15 Feb. 1897. MS. Pearson Papers. University College London, London.

—. *Memories of My Life.* New York: Dutton, 1909. Print.

—. *Natural Inheritance.* London: Macmillan, 1889. Print.

—. "Regression towards Mediocrity in Hereditary Stature." *Journal of the Anthropological Institute* 15 (1886): 246–53. Print.

"Gene." *Oxford English Dictionary Online*. Web. 27 Mar. 2006.

Rev. of *The Genetical Theory of Natural Selection. Quarterly Review of Biology* 6.1 (1931): 100. Print.

Ghislen, Michael T. *The Triumph of the Darwinian Method.* Berkeley: U of California P, 1969. Print.

Gieryn, Thomas. *Cultural Boundaries of Science: Credibility on the Line.* Chicago: U of Chicago P, 1999. Print.

Gillham, Nicholas Wright. *A Life of Sir Francis Galton.* Oxford: Oxford UP, 2001. Print.

Gliboff, Sander. *H.G. Bronn, Ernst Haeckel, and the Origins of German Darwinism: A Study of Translation and Transformation.* Cambridge: MIT, 2008. Print.

Gonick, Larry, and Woollcott Smith. *The Cartoon Guide to Statistics.* New York: Harper Collins, 1993. Print.

Gray, Asa. "Review of Darwin's Theory of the Origin of Species by Means of Natural Selection." *American Journal of Science and Arts* 29.86 (1860): 153–185. Print.

Gross, A. "The Origin of the Species: Evolutionary Taxonomy as an Example of the Rhetoric of Science." *The Rhetorical Turn: Invention and Persuasion in the Conduct of Inquiry.* Ed. Herbert Simons. Chicago: U of Chicago P, 1990. Print.

Gross, Alan, Joseph Harmon, and Michael Reidy. *Communicating Science: The Scientific Article from the 17th Century to the Present.* Oxford: Oxford UP, 2002. Print.

Hacking, Ian. *The Taming of Chance.* Cambridge: Cambridge UP, 1990. Print.

Haldane, J.B.S. "Forty Years of Genetics." *Background to Modern Science.* Joseph Needham and Walter Pagel Eds. Cambridge: Cambridge UP, 1938. Print.

—. Rev. of *The Genetical Theory of Natural Selection. The Mathematical Gazette* 15.215 (1931): 474–75. Print.

—. "Some Statistical Problems Arising in Genetics." *Journal of the Royal Statistical Society. Series B (Methodological)* 11.1 (1949): 1–14. Print.
Hartl, Daniel and Elizabeth Jones. "Molecular Evolution and Population Genetics." *Genetics*. Boston: Jones and Bartlett, 2009. Print.
Harris, George. Rev. of *Hereditary Genius. Journal of Anthropology* 1.1 (1870): 56–65. Print.
Harvey, Rosemary. *William Bateson and the Emergence of Genetics: A Biography in Five Volumes*. Vol 2. Norwich: The John Innes Center, 2000. TS.
Henig, Robin Marantz. *The Monk in the Garden*. Boston: Houghton Mifflin, 2000. Print.
Rev. of *Hereditary Genius. Atlantic Monthly* 25.152 (1870): 753–56. Print.
—. *Galaxy* 9.3 (1870): 424. Print.
—. *The Westminster Review* 83.43 (1870): 144–45. Print.
Herivel, John. Introduction. *Philosophy of the Inductive Sciences*. By William Whewell. 2nd ed. 2 vols. London: Parker, 1847. New York: Johnson Reprint, 1967. Print.
Herschel, John. *A Preliminary Discourse on the Study of Natural Philosophy*. 1830. New York: Johnson Reprint, 1966. Print.
Hill, Austin Bradford. Rev. of *The Genetical Theory of Natural Selection. Journal of the Royal Statistical Society* 94.1 (1931): 98–100. Print.
Hooker, Joseph. *The Botany of the Antarctic Voyage of H.M. Discovery Ships* Erebus *and* Terror *in the Years 1839–1843: Under the Command of Captain Sir James Clark Ross, Sir Joseph Dalton Hooker*. 3 vols. London: Reeve, 1844–1863. Print.
—."On the Origination and Distribution of Species—Introductory Essay to the Flora of Tasmania." *American Journal of Science and Arts* 29.85 (1860): 1–25. Print.
—. "To Charles Darwin." 14 Mar. 1858. *Darwin Correspondence Project*. American Council of Learned Societies and Cambridge University. Web. 14 July 2008.
Hopkins, William. "Physical Theories of the Phenomena of Life." *Fraser's Magazine* 62 (1860): 74–90. Print.
Houlette, Forrest. *Nineteenth Century Rhetoric: An Enumerative Bibliography*. New York: Garland, 1989. Print.
Hull, David. *Darwin and His Critics: The Reception of Darwin's Theory of Evolution by the Scientific Community*. Cambridge: Harvard UP, 1973. Print.
"Human Evolutionary Genetics." Wikipedia. Web.14 Dec. 2009.
Humboldt, Alexander. *Aspect of Nature in Different Lands and Different Climates; with Scientific Elucidation*. Vol. 2. London: Murray, 1849. Print.
Humboldt, Alexander and Aime Bonpland. *Personal Narrative of Travels to the Equinocturnal Regions of the New Continent during the Years 1799–1804*. Trans. Helen Williams. 7 vols. London, 1814–1829. Print.

Hunt, Bruce. "Rigorous Discipline: Oliver Heaviside Versus the Mathematicians." *The Literary Structure of Scientific Argument.* Ed. Peter Dear. Philadelphia: U of Pennsylvania P, 1991. Print.

Iltis, Hugo. *Life of Mendel.* Trans. Eden and Cedar Paul. New York: Norton, 1932. Print.

Jones, Harold Ellis. "Homogamy in Intellectual Abilities." *The American Journal of Sociology* 35 (1929): 369–382. Print.

Jungnickel, Christa and Russell McCormick. *Cavendish.* Philadelphia: The American Philosophical Society, 1996. Print.

Kant, Immanuel. *Metaphysical Foundations of Natural Science.* Ed. Michael Friedman. Cambridge: Cambridge UP, 2004. Print.

Kelley, Truman. "Measures of Correlations Determined from Groups of Varying Homogeneity." *Journal of the American Statistical Association* 20 (1925): 512-21. Print.

Kemsley, Rachael, ed. "Herschel, Sir John Fredrick William (1792–1871)." Sept. 2002. *AIM 25.* Web. 15 Jul. 2007.

Knorr-Cetina, Karin. *Epistemic Cultures: How the Sciences Make Knowledge.* Cambridge: Harvard UP, 1999. Print.

Kölreuter, Joseph Gottlieb. *Vorläufige Nachricht von einigen das Geschlecht der Pflanzen betreffenden Versuchen und Beobachtungen nebst Fortsetzungen 1, 2, und, 3* (Preliminary Report about some Observations and Experiments Regarding the Gender of Plants with Continuations 1, 2, and 3). 4 vols. Leipzig, 1893.

Kuhn, Thomas. *The Structure of Scientific Revolutions.* 3rd ed. Chicago: U Chicago P, 1996. Print.

Lamarck, Jean Baptist. *Philosophie Zoologique.* Paris, 1809. *Google Book Search.* Web. 22 Dec. 2009.

Latour, Bruno, and Steve Woolgar. *Laboratory Life.* Princeton: Princeton UP, 1986. Print.

Lewontin, Richard. "Theoretical Population Genetics in the Evolutionary Synthesis." *The Evolutionary Synthesis: Perspectives on the Unification of Biology.* Eds. Ernst Mayr and William Provine. Harvard UP, 1980. Print.

Lyell, Charles. *Principles of Geology: Being an Attempt to Explain the Former Changes of the Earth's Surface.* 5th ed. Vol. 3. London, 1837. *Google Book Search.* Web. 13 June 2008.

Lynch, Michael. *The Origins of Genome Architecture.* Sunderland: Sinauer, 2007. Print.

MacKenzie, Donald A. *Statistics in Britain 1865–1930: The Social Construction of Scientific Knowledge.* Edinburgh: Edinburgh UP, 1981. Print.

Magnello, Eileen M. "Karl Pearson's Gresham Lectures: W.F.R. Weldon, Speciation and the Origins of Pearsonian Statistics." *British Journal for the History of Science* 29 (1996): 43–63. Print.

Mather, Kenneth. Rev. of *Commentary on R.A. Fisher's Paper on The Correlation between Relatives on the Supposition of Mendelian Inheritance. Population Studies* 20.3 (1967): 372–73. Print.

Maxwell, James Clerk. *A Discourse on Molecules*. Bradford, 1873. Print.

Mayr, Ernst. *The Growth of Biological Thought*. Cambridge: Harvard UP, 1982. Print.

Mendel, Gregor. *Experiments in Plant Hybridization*. Cambridge: Harvard UP, 1965. Print.

—. *Versuche über Plfanzen-Hybriden*. Eds. J. Cramer and H.K. Swann. New York: Hafner, 1960. Print.

Rev. of *Mendel's Principles of Heredity. Athenaeum* 4 Sep. 1909. Print.

—. *Daily Telegraph* 3 June 1909. Print.

—. *Lancet* 173.4473 (1909): 1461–62. Print.

—. *Manchester Courier* 28 May 1909. Print.

—. *Nation* 16 Sep. 1909. Print.

—. *Saturday Review* 22 Jan. 1910. Print.

Mill, John Stuart. *System of Logic, Ratiocinative and Inductive: Being a Connected View of the Principles of Evidence and Methods of Scientific Investigation*. 1886. Whitefish: Kessinger, 2000. Print.

Miller, Carolyn. "The Presumption of Expertise: The Role of Ethos in Risk Analysis." *Configurations* 11.2 (2003): 163–202. Print.

Miller, Carolyn and S. Michael Halloran. "Reading Darwin, Reading Nature: Or, on the Ethos of Historical Science." *Understanding Scientific Prose*. Ed. Jack Seltzer. Madison: U of Wisconsin P, 1993. 106–26. Print.

"Monocotyledon." Wikipedia. Web. July 22, 2008.

Moss, Jean Dietz. *Novelties in the Heavens*. Chicago: U of Chicago P, 1993. Print.

"Mr. Galton on Natural Inheritance." Rev. of *Natural Inheritance. Times* 13 Aug. 1889: 3. Print.

Myers, Greg. *Writing Biology: Texts and the Social Construction of Scientific Knowledge*. Madison, U of Wisconsin P, 1990. Print.

Nägeli, Carl von "Die Bastardbildung im Pflanzenreich," *Botanische Mittheilungen* 2 (1865): 187–235. Print.

National Science Foundation. "Water Lily May Provide a 'Missing Link' in the Evolution of Flowering Plants." *Science Daily* 31 Jan. 2002. Web. 14 Dec. 2009.

Rev. *Natural Inheritance. Nation* 5 Sep. 1889: 196–98. Print.

—. *Science* 12.322 (1889): 266–67. Print.

Norton, Bernard and E. S. Pearson. "A Note on the Background to, and Refereeing of, R. A. Fisher's 1918 Paper 'On the Correlation between Relatives on the Supposition of Mendelian Inheritance.' *Notes and Records of the Royal Society of London* 31.1 (1976): 151–162. Print.

Olby, Robert. *The Origins of Mendelism*. New York: Schocken, 1966. Print.

Olson, Steve. *Mapping Human History.* New York: Houghton Mifflin, 2002. Print.

Orel, Vitezslav. *Mendel.* Oxford: Oxford UP, 1984. Print.

Parshall, Karen Hunger. "Varieties as Incipient Species: Darwin's Numerical Analysis." *Journal of the History of Biology* 15 (1982): 191–124. Print.

Partridge, Michael. Introduction. *A Preliminary Discourse on the Study of Natural Philosophy.* 1830. By William Herschel. New York: Johnson Reprint, 1966. Print.

Patriarca, Silvana. *Numbers and Nationhood: Writing Statistics in Nineteenth Century Italy.* Cambridge: Cambridge UP, 1996. Print.

Paul, Danette. "In Citing Chaos: A Study of a Rhetorical Use of Citations." *Journal of Business and Technical Communication* (2000):185–222. Print.

Paul, Dannette, Davida Charney, and Aimee Kendall. "Moving Beyond the Moment: Reception Studies in the Rhetoric of Science." *Journal of Technical and Business Communication* 15.3 (2001): 372–399. Print.

Pearl, Raymond. *Modes of Research in Genetics.* New York: Macmillan, 1915. Print.

Pearson, E. S. "Studies in the History of Probability and Statistics. XX: Some Early Correspondence between W.S. Gossett, R.A. Fisher, and Karl Pearson, with Notes and Comments." *Biometrika* 55.3 (1968): 445–457. Print.

Pearson, Karl. "On a Certain Atomic Hypothesis." *Transactions of the Cambridge Philosophical Society* 14 (1889): 71–120. Print.

—."To Francis Galton." 12 Feb. 1897. MS. Karl Pearson Papers. University College London, London.

—. *The Grammar of Science.* London, 1892. Print.

—. *The Grammar of Science.* London, 1900. Print.

—."Mathematical Contributions to the Theory of Evolution [abstract]." *Proceedings of the Royal Society London* 45 (1893): 329–333. Print.

—. "Mathematical Contributions to the Theory of Evolution. IX. On the Principle of Homotyposis and Its Relation to Heredity, to the Variability of the Individual, and to that of the Race. Part I. Homotypos in the Vegetable Kingdom." *Philosophical Transactions of the Royal Society of London. Series A, Containing Papers of a Mathematical or Physical Character* 197 (1901): 285–379. Print.

—. "Mathematical Contributions to the Theory of Evolution. XII. On a Generalized Theory of Alternative Inheritance with Special Reference to Mendel's Laws." *Philosophical Transactions of the Royal Society London. Series A, Containing Papers of a Mathematical or Physical Character* 203 (1904): 53- 86. Print.

—. "The Scope of Biometrika." *Biometrika* 1.1 (1901): 1–2. Print.

—. "The Spirit of Biometrika." *Biometrika* 1.1 (1901): 3–6. Print.

Perelman, Chaim, and Lucie Olbrechts-Tyteca. *The New Rhetoric.* Notre Dame: U of Notre Dame P, 1971. Print.

Plutynski, Anya. "Explanation in Classical Population Genetics." *Philosophy of Science* 71 (2004): 1201–1214. Print.

Prelli, Lawrence. *A Rhetoric of Science: Inventing Scientific Discourse.* Columbia: U of South Carolina P, 1989. Print.

—. "The Rhetorical Construction of Scientific Ethos." *Landmark Essays on Rhetoric of Science.* Ed. Randy Allen Harris. Mahwah: Erlbaum, 1997. Print.

"Population Genetics." *Oxford English Dictionary Online.* Web. 2 Nov. 2009.

Porter, Theodore M. *Karl Pearson: The Scientific Life in a Statistical Age.* Princeton: Princeton UP, 2004. Print.

"Problems of Life." Rev. of *The Genetical Theory of Natural Selection. Spectator* 24 May 1930, 870. Print.

Provine, William. *The Origins of Theoretical Population Genetics.* 1971. Chicago: U of Chicago P, 2001. Print.

Punnett, R. C., "Genetics, Mathematics, and Natural Selection." Rev. of *The Genetical Theory of Natural Selection. Nature* 126 (1930): 595–97. Print.

—. ed. *Scientific Papers of William Bateson.* 2 vols. Cambridge: Cambridge UP, 1928. Print.

Reyes, G. M. "The Rhetoric in Mathematics: Newton, Leibnitz, the Calculus, and the Rhetorical Force of the Infinitesimal." *Quarterly Journal Speech* 90.2 (2004): 163–88. Print.

Rhetorica Ad Herennium. Trans. Harry Caplan. Cambridge: Harvard UP, 1999. Print.

Richards, Joan. *Mathematical Visions: The Pursuit of Geometry in Victorian England.* Boston: Academic P, 1988. Print.

Roberts, H. F. *Hybridization before Mendel.* Princeton: Princeton UP, 1929. Print.

Schweber, Silvan. "Darwin and the Political Economists: Divergence of Character." *Journal of the History of Biology* 13.2 (1980): 195–289. Print.

"The Scientific Treatment of Statistics." Rev. of *Natural Inheritance. Spectator* 20 July 1889: 83–84. Print.

Shull, George H. "Heredity as an Exact Science." *Botanical Gazette* 50.3 (1910): 226- 229. Print.

Snyder, Laura. *Reforming Philosophy: A Victorian Debate on Science and Society.* Chicago: U of Chicago P, 2006. Print.

Stanley, Hiram. "Mr. Galton on Natural Inheritance." Rev. of *Natural Inheritance. Nature* 40.1044 (1889): 642–43. Print.

Stern, Curt, and Eva Sherwood eds. *The Origin of Genetics: A Mendel Source Book.* San Francisco: Freeman, 1966. Print.

Stigler, Stephen. *The History of Statistics: The Measurement of Uncertainty before 1900.* Cambridge: Harvard UP, 1986. Print.

Swales, John. *Genre Analysis: English in Academic and Research Settings.* Cambridge: Cambridge UP, 1990. Print.

Sykes, Brian. *The Seven Daughters of Eve.* New York: Norton, 2001. Print.

Templeton, Alan. Telephone interview. 9 Dec. 2009.

Thompson, Herbert. "On Correlation of Certain External Parts of Paloemon serratus." *Proceedings of the Royal Society of London* 55 (1894): 234–40. Web. 21 June 2010.

Toulmin, Stephen. *An Introduction to Reasoning.* 2nd ed. New York: Macmillan, 1984. Print.

—. *The Uses of Argument.* Updated ed. Cambridge: Cambridge UP, 2003. Print.

Venn, John. Rev. of *Natural Inheritance. Mind* 14.55 (1889): 414-20. Print.

Verhandlungen des naturforschenden Vereines in Brunn 4 (1866): Front Matter. Print.

Watson, H. C. "To Charles Darwin." 19 Nov. 1854. *Darwin Correspondence Project.* American Council of Learned Societies and Cambridge University. Web. 14 July 2008.

—. "To Charles Darwin." 17 Aug. 1855. *Darwin Correspondence Project.* American Council of Learned Societies and Cambridge University. Web. 14 July 2008.

—. "To Charles Darwin." 8 Nov. 1855. *Darwin Correspondence Project.* American Council of Learned Societies and Cambridge University. Web. 20 July 2008.

Weldon, W. F. R. "To Francis Galton." 27 Jan. 1895. MS. Galton Papers. University College London, London.

—. "To Francis Galton." 11 Feb. 1895. MS. Galton Papers. University College London, London.

—. "To Francis Galton." 17 Nov. 1896. MS. Galton Papers. University College London, London.

—. "To Francis Galton." 26 Nov. 1896. MS. Galton Papers. University College London, London.

—. "To Francis Galton." 28 Nov. 1899. MS. Galton Papers. University College London, London.

—. "To Karl Pearson." 23 Dec. 1905. MS. Pearson Papers. University College London, London.

—. "Mendel's Laws of Alternative Inheritance in Peas." *Biometrika* 1.2 (1902): 228-254. Print.

—. "The Variations Occurring in Certain Decapod Crustaceans. I. *Cragon vulgaris.*" *Proceedings of the Royal Society* 47 (1889–90): 445–453. Print.

Wells, David, perf. *Journey of Man: The Story of the Human Species.* Dir. Clive Maltby. PBS, 2003. DVD.

Westergaard, Harald. *Contributions to the History of Statistics.* New York: Agathon P, 1968. Print.

Whewell, William. *History of the Inductive Sciences.* 3rd ed. 3 vols. 1857. London: Cass, 1967. Print.

—. *Philosophy of the Inductive Sciences.* 2nd ed. 2 vols. London: Parker, 1847. New York: Johnson Reprint, 1967. Print.

Wright, Seawall. Rev. of *The Genetical Theory of Natural Selection. Journal of Heredity* 21.8 (1930): 349–56. Print.

Yule, Udny. "Mendel's Laws and their Probable Relations to Interracial Heredity." *The New Phytologist* 1.9 (1902): 193–207, 222–238. Print.

Index

About the Author

James Wynn is an Associate Professor of English and Rhetoric at Carnegie Mellon University. He has published articles on rhetoric, mathematics, and science in *Rhetorica*, *Written Communication*, and *19th Century Prose.* His recent interests have been in rhetoric, science, mathematics and public policy with a focus on nuclear power. He is a founder and current director of the Pittsburgh Consortium for Rhetoric and Discourse Studies.

www.ingramcontent.com/pod-product-compliance
Lightning Source LLC
Chambersburg PA
CBHW021552210326
41599CB00010B/413

* 9 7 8 1 6 0 2 3 5 2 1 6 2 *